Series in BioEngineering

The Series in Bioengineering serves as an information source for a professional audience in science and technology as well as for advanced students. It covers all applications of the physical sciences and technology to medicine and the life sciences. Its scope ranges from bioengineering, biomedical and clinical engineering to biophysics, biomechanics, biomaterials, and bioinformatics.

Indexed by WTI Frankfurt eG, zbMATH.

R. Prasad

Understanding Medical Devices

The Science Behind the Diagnosis

Springer

R. Prasad
Emeritus Professor
Aligarh Muslim University
Aligarh, Uttar Pradesh, India

ISSN 2196-8861 ISSN 2196-887X (electronic)
Series in BioEngineering
ISBN 978-3-031-88629-4 ISBN 978-3-031-88630-0 (eBook)
https://doi.org/10.1007/978-3-031-88630-0

This Springer imprint is published by the registered company Springer Nature Switzerland AG
The registered company address is: Gewerbestrasse 11, 6330 Cham, Switzerland

Dedicated to my wife Sushma Mathur
Who inspired me to write this book
(12 November 1950–24 April 2023)

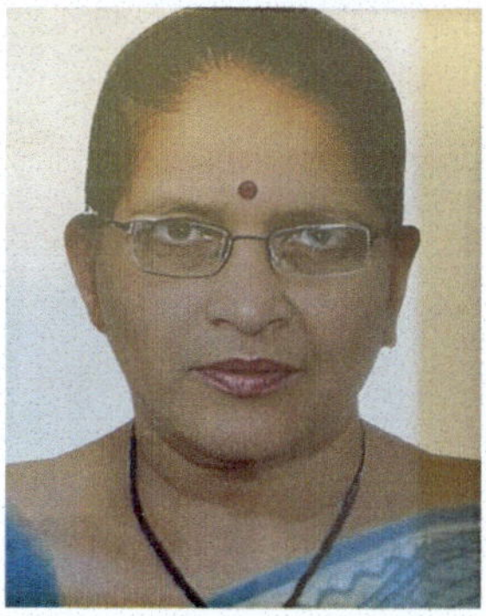

Beautiful lady fought bravely with cancer for three years

Preface

Recently, I had to pass a considerable time in different hospitals in Sydney, Australia, as my wife was hospitalized for the treatment of cancer. During this period, she had to undergo different diagnostic tests, some in her hospital room and others in specific laboratories where I used to accompany her. I always used to ask the doctor/nurse/technician performing the test if they know how the gadget/instrument they are using works and how the result of the test is displayed? In most cases, without a single exception, the answer was NO. Senior doctors said that they do not really know the working of the diagnostic tools; however, they know how to interpret the result. My next question was, would they like to know the working of the instruments? And their emphatic answer was YES. Some, mostly young doctors, expressed the view that they are not required to know the working of the instruments and questioned back what will they gain by knowing the working of the instruments? My reply was, firstly they will not be harmed in any way if they know how a gadget works and more importantly, they may suggest improvements/modifications that may further improve the analyzing power of the instrument if they understand how the instrument works. Then I narrated some instances where doctors/technicians conversant with the working of an instrument suggested modifications that brought the gadget in its present form. This, to some extent, convinced them.

I then raised the same question with my friends, some of whom are in the field of education, some in business and in other walks of life. They all appeared eager to know, for example, how a strip thermometer gives the temperature in digital form, or how a glucometer gives the blood glucose level or how a gun thermometer tells body temperature from a distance. Some physics professors show exceptional interest in knowing the working of diagnostic tools they frequently use in their daily life.

In light of my above experiences, I decided to write, in the first phase, on the working of ten instruments/gadgets frequently used to diagnose ailments/medical condition of our body parts.

The book is written in simple language, avoiding derivations and complex mathematics as far as possible. Attempt has been made to explain the text by giving suitable examples from our daily life and incorporating explanatory figures. Use of some scientific and medical terminology is unavoidable, but it is explained immediately by enclosing within brackets the meaning of the term in simple language.

Most of the modern diagnostic gadgets are digital and have digital display screens, and the data recorded by these instruments can be digitally shared with others at far-off places using an Internet network. To understand and appreciate the versatility of these gadgets, it is required to give a short description of the process of digitization and how it is achieved. Since digitization and many other functions involved in the working of these instruments are based on the processing of electronic pulses, topics like electric pulse, transducers, analog to digital (A/D) and (D/A) converters, binary and other number systems, digital display screens, electromagnetic radiations, etc., that are required to understand the working of these tools are all discussed in the section "Basics" of the document. The "Basics" is one of the most important section of the document. A reader is advised to go through the section "Basics" before proceeding any further.

The book contains 10 chapters, and a section "Basics" that deal with the basic information about binary systems, digital and analog signals, digital to analog and analog to digital conversions, digital display screens, electromagnetic radiations and their classification, etc. A basic diagnostic gadget, the stethoscope, which has become a doctor identifier, is discussed in Chap. 1. The working of the instrument, essentially based on the principles of multiple reflection and resonance of sound waves, has been explained by giving examples from our daily life. A raised temperature of the body is invariably taken as a warning sign of an unwell body. Heat generation and temperature control mechanisms in the human body are outlined in Chap. 2. Detailed working of different types of thermometers, including the gun-type remote sensing thermometer, are discussed in this chapter. Chapter 3 gives details of the working of a pulse oximeter, an instrument used to monitor the level of oxygen saturation in blood and the pulse rate. This gadget became very popular during the period of COVID-19 pandemic because the disease rapidly reduced the oxygen level in patients; therefore, it was required to almost continuously monitor it. Sphygmomanometer, an instrument used to measure blood pressure, is discussed in Chap. 4. What is meant by the term blood pressure, how it develops and monitored, the significance of high and low blood pressures and risks involved, etc., are all discussed in detail in Chap. 4. Basic technique of X-ray imaging of body parts is given in Chap. 5. Production of X-rays, their important properties, X-ray sources, X-ray detectors, image intensifiers, image parameters, etc., are described in sufficient detail in this chapter. Computed tomography (CT scan) technique, often used to identify the cause of some ailments is given in Chap. 6. The technique uses the X-ray imaging of the given object from different angles and generates a detailed image of the object by the computer synthesis of several complementary X-ray images. Magnetic resonance imaging (MRI) is a technique where detailed images of almost every internal structure in the body, including the organs, bones, muscles, and blood vessels may be generated using magnetic field and radio waves; it is discussed in Chap. 7. Positron emission tomography (PET) imaging is a medical imaging procedure that provides unique information on the working of different body parts, systems and organs. It is based on identifying locations of high metabolic activity in the body using radioactive isotopes that emit positrons. Details of the working of the PET system are elaborated in Chap. 8. Chapter 9 provides the details of the non-invasive and one of the safest technologies of ultrasound imaging. Doppler ultrasound images may record the flow pattern of blood in different body parts. The working of the glucometer, an instrument to measure the concentration of sugar in blood, is explained in Chap. 10. Different types of glucometers and their relative merits/demerits are also discussed in this chapter.

Some readers may find repetitions in the book. It is deliberate. Because of the mixed nature of readership, it is sometimes required to repeat a specific concept/principle/expression at two or three places to strengthen its understanding. Though an attempt has been made to avoid scientific and medical terminology, it is unavoidable in a book like this. However, the meaning of the term is explained in simple language, if used.

Suggestions from readers to further improve the book are welcome. I am always available on my E-mail: rpm166@rediffmail.com.

Kellyville, NSW, Australia R. Prasad

Acknowledgements I wish to thank from the depth of my heart my late wife Sushma Mathur who inspired me to write this book. While bravely fighting cancer for almost three years and at her deathbed, knowing well that she is not going to live further, she asked me not to waste time and brood over her misery, instead do some creative work. She used to watch doctors in hospital taking different types of data using diagnostic machines/gadgets or scanners. While with her, I used to ask these doctors do they know how the gadget/scanner they used works and takes the data? Most of the time doctors replied in negative, adding that they know how to interpret the result. Sushma asked me to write a book explaining the science/physics behind the working of these gadgets. She was sure that I could do it. As such the book belongs to her. Thank you, Sushma.

I discussed the subject matter of the book with many colleagues, including Prof. Kuldeep Paliwal, Professor of Electronics at the Griffith University, Brisbane, Australia. I wish to put on record my sincere thanks to Prof. Paliwal who not only made many useful suggestions but took the trouble of reading and correcting some chapters of the book. A big thank you Prof. Paliwal.

My grandchildren Antra, Ansh, Aaniya and Aaliya deserve thanks; Antra and Ansh for helping me in working with my new laptop and the new platform. Problems associated with my drive, on which I wasted hours, were resolved by Antra and Ansh in seconds. Thank you, Antra and Ansh. Aaniya and Aaliya have been great mood elevators, while I was depressed or tired. Thank you both.

Kellyville, NSW, Australia R. Prasad

Competing Interests The author has no competing interests to declare that are relevant to the content of this manuscript.

Basics

Electromagnetic Radiations

When we stand in the Sun for a while we feel warmth. We also experience heat when we stand in front of a hot object like a bucket of hot water. Both these happenings show that heat, which is a kind of energy, reaches us from a hot object (Sun is a hot object) passing through the intervening space. The fact that there is a vast vacuum (region without any medium like air, etc.) between the Sun and the earth yet heat energy from the Sun reaches the man standing in the sunlight on the earth, proves that heat energy may travel through vacuum. Detailed investigations into the nature of heat radiation have revealed that all hot bodies emit radiation that carries energy with them. These radiations are called ***Electromagnetic (EM) Waves or Radiations***. The name electromagnetic comes from the fact that these radiations are made up of electric and magnetic fields in two mutually perpendicular directions as shown in Fig. 1. As shown in the figure, the intensities of both the electric and the magnetic fields vary with time with a given repetition rate.

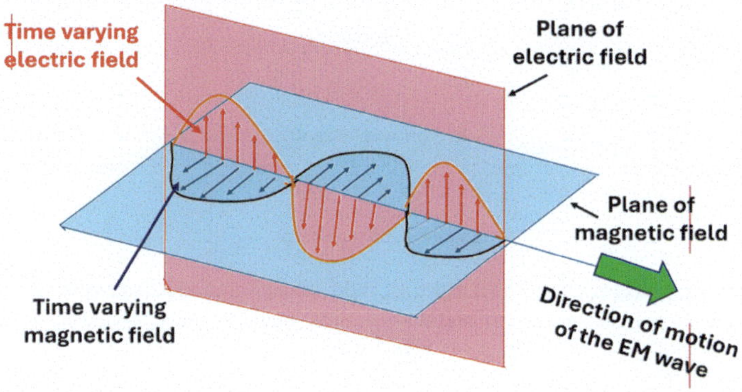

Fig. 1 Rough sketch of an electromagnetic wave

The figure shows an electromagnetic wave made up of an electric field contained in the vertical plane (along the Y-direction) and the magnetic field contained in the horizontal plane (along X-direction). The magnitude and the direction of both the electrical and the magnetic fields in their respective planes keep changing regularly with time, as indicated by the size of the red and the blue arrows. The direction of propagation of the EM wave, shown by the thick green arrow in the figure, is along the Z-axis; it may be noted that the direction of propagation of the EM wave is perpendicular both the vertical plane of the electric field and the horizontal plane of the magnetic field.

Generation of EM waves: Suppose there is a ***stationary electric charge*** of magnitude **Q** placed at some point P in space. It is found that such a stationary electric charge produces an electric field **ε** in the space surrounding the charge. Next suppose that the electric charge **Q** moves with a constant speed, say $\mathcal{V}$, then this uniform motion of the charge constitutes an electric current and produces both the electric field **ε** and the magnetic field **H** in the space surrounding the current. Consider the special case when the charge Q moves with a non-uniform speed, which is the charge either gets accelerated or decelerated (negatively accelerated). It can be shown that an accelerated (or decelerated) charge is equivalent to a current that changes in magnitude. It is found that an accelerated electric charge or a current element of changing intensity radiates the EM wave and loses energy. In short, EM waves are produced by accelerated (or decelerated) electric charges or by the currents that vary in magnitude. Figure 2 shows the three situations when charge is stationary, moving with uniform speed and when it is accelerated.

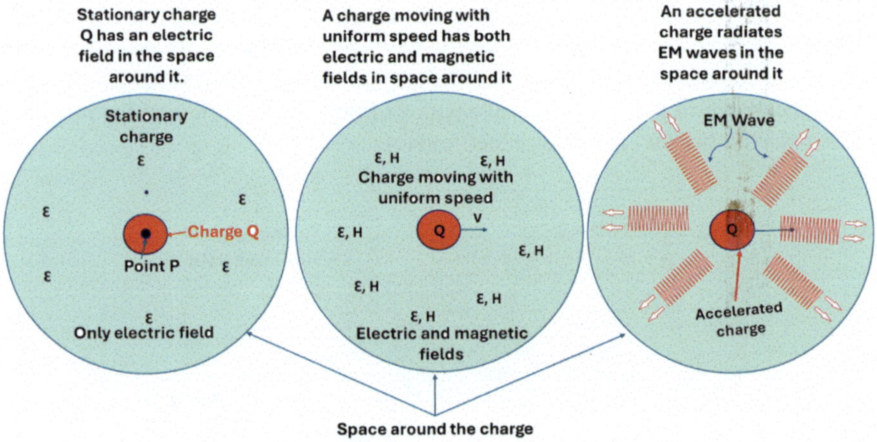

Fig. 2 A stationary electric charge has an electric field in space around it. An electric charge moving at uniform speed has both the electric and the magnetic fields in space around it. An accelerated electric charge radiates energy in the form of EM waves

Why a Hot Body Radiates EM Waves

As shown in Fig. 2, an accelerated (or decelerated) electric charge or current element with changing magnitude radiates electromagnetic waves in the space around it. All bodies and objects are made up of molecules and atoms, and each atom of the object has a positively charged nucleus along with negatively charged electrons. These charged particles (nucleus and electron) are set in accelerated motion when the body is heated, and therefore, hot objects emit EM waves.

What is an EM Wave

An electromagnetic wave has an electric and a magnetic field confined in two mutually perpendicular planes. If it is assumed that the electric field of the EM wave is confined in the Y-plane, then the magnetic field of the wave will be confined in the X-plane. This combination of electric and magnetic fields in mutually perpendicular planes moves away from the source (hot object) in the Z-direction, which is perpendicular to both the X- and the Y-planes. Further, the magnitudes of the electric and the magnetic fields change rapidly with time as the wave moves ahead away from the source, the hot object. The magnitude of both the electric and the magnetic fields in an EM wave first increases from the zero value to the maximum value in one direction, then decreases to zero value and again increases to the maximum value in the opposite direction from where it decreases to the zero value again. And this variation in intensity of the field goes on. Thus, both the electric and the magnetic fields undergo periodic variation in their magnitudes in their specific planes which are perpendicular to each other as the EM wave moves. A rough sketch of the time variation of the electric field of an EM wave is shown in Fig. 3.

As shown in the figure, the time variation of field intensity may be segmented into several identical parts, each of a ***complete cycle*** of intensity variation. The time taken in one complete cycle is called the ***time period*** of the wave and is often denoted by T. The time period T is measured in units of second, denoted by **s**. The reciprocal of the time period T is called the ***frequency*** of the wave, denoted by the Greek letter **γ**;

$$\gamma = 1/\mathrm{T} \tag{1}$$

Since frequency is reciprocal of time period, **γ** is measured in unit that is reciprocal of time, called ***Hertz*** and denoted by ***Hz***. It may be noted that the frequency of intensity variation for both the electric and the magnetic fields of a given EM wave is the same.

Frequency γ and the speed of travel **V** are the two characteristic parameters of an EM wave, which are independent of each other. The frequency **γ** depends on the source of the EM wave (the hot body that generates the EM wave) while the speed **V** depends on the medium through which the wave is moving. If the EM wave is

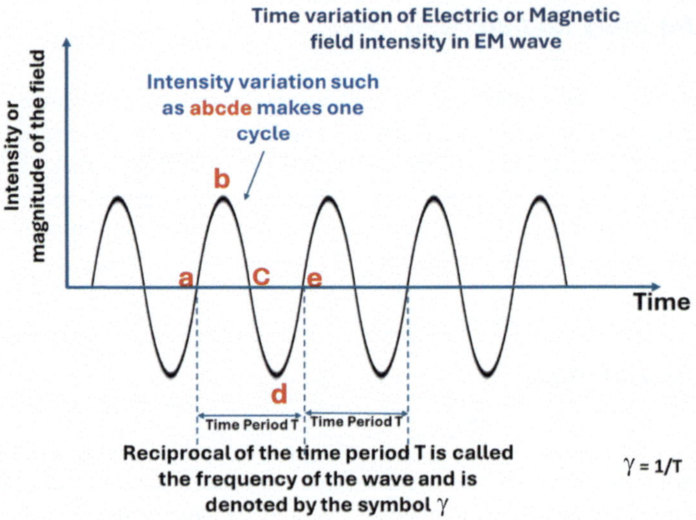

Fig. 3 Typical graph for the time variation of electric or magnetic field intensity in an electromagnetic wave

moving in vacuum, then the speed of the wave, irrespective of its frequency, is taken as 3×10^8 m/s ($=3 \times 10^{10}$ cm/s) and is denoted by the small English alphabet **c**. The speed of EM waves in air (at normal atmospheric pressure and temperature) is also taken to be the same as in vacuum, i.e., **c**. However, in denser medium the EM wave travels with smaller speeds; for example, the speed of EM waves in water is 2.26×10^{10} cm/s. Since the frequency of an EM wave is the property of its source, it does not change no matter in which medium the EM wave travels; however, the speed of the electromagnetic wave depends on the medium.

Another parameter, called ***wavelength*** and denoted by the Greek letter **λ**, may also be associated with a wave. Wavelength depends both on the frequency γ (property that depends only on source) and the speed **V** (that depends on the medium) through the following relation.

$$(Frequency) \times (Wavelength) = Speed \quad \text{or} \quad \gamma \times \lambda = \mathrm{V} \tag{2}$$

Wavelength **λ** may be measured in several different units that represent distance. For example, wavelength may be measured in meter (**m**), centimeter (**cm, 10^{-2} m**), millimeter (**mm, 10^{-3} m**), micrometer (**μm, 10^{-6} m**), nanometers (**nm, 10^{-9} m**), etc. Equation 2 tells that the wavelength **λ** and the frequency ***γ*** have inverse relationship, if one increases, the other decreases. A large value of frequency corresponds to a small value of wavelength and vice versa.

As already mentioned, EM waves are generated when electric charges undergo acceleration or deceleration. It may be said that the frequency γ of the emitted EM wave is related to the magnitude of acceleration/deceleration of the electric charge. Since the tiniest bodies have a very large number of atoms, each atom has a positively charged nucleus (an electric charge) and several electrons; there are always many electric charges held inside a given body. Now if such a body is heated, these charges inside the body get accelerated; however, the acceleration of different charges has different values. As a result, ***a hot body emits EM waves of many different frequencies depending on the temperature of the hot body.*** The higher the temperature of the body, the larger number of EM waves of higher frequencies are emitted.

Dual Nature of EM Waves

Experiments have indicated that EM waves in some experiments behave as waves, showing properties associated only with waves, while in some other experiments they behave as if they are made up of particles. However, the important point is that in experiments where they behave as particles, they do not exhibit wave properties, and similarly, in experiments where EM waves behave as waves, they do not show their particle nature. To reconcile this dual nature of EM waves, often one gives the example of a coin, when one looks at the head of the coin, the other face, called the tail is not visible and similarly while viewing the tail side of the coin the head is not visible. The constituent particle of an EM wave is called ***PHOTON***. It is taken that an EM wave of frequency γ is made up of photons, each of which has an energy of ($\hbar\gamma$) eV. The number of photons per unit volume in an EM wave determines the intensity of the wave. In a way, photons in an EM wave may be considered as a packet each of energy ($\hbar\gamma$) eV that are emitted by the source. The symbol $\hbar$ stands for a constant, called ***rationalized Planck's constant*** with the value **6.58 x 10^{-16} eV s**; here eV, called electron volt, is a unit of energy and s is the symbol for second. 1 eV = 1.6 × 10^{-19} J (Joule) = 1.6 × 10^{-12} erg. Since energy ε of a photon of frequency γ is equal to $\hbar\gamma$, i.e.,

$$\varepsilon = \hbar\gamma \tag{3}$$

a photon of high frequency (or of smaller wavelength) will carry more energy as compared to the photon that has smaller frequency (or larger wavelength).

Frequency (or Wavelength) Spectrum of EM Waves

Figure 4 shows the complete spectrum of electromagnetic waves. As may be seen in the middle part of the figure, the full spectrum may be divided into the following six broad frequency or wavelength components.

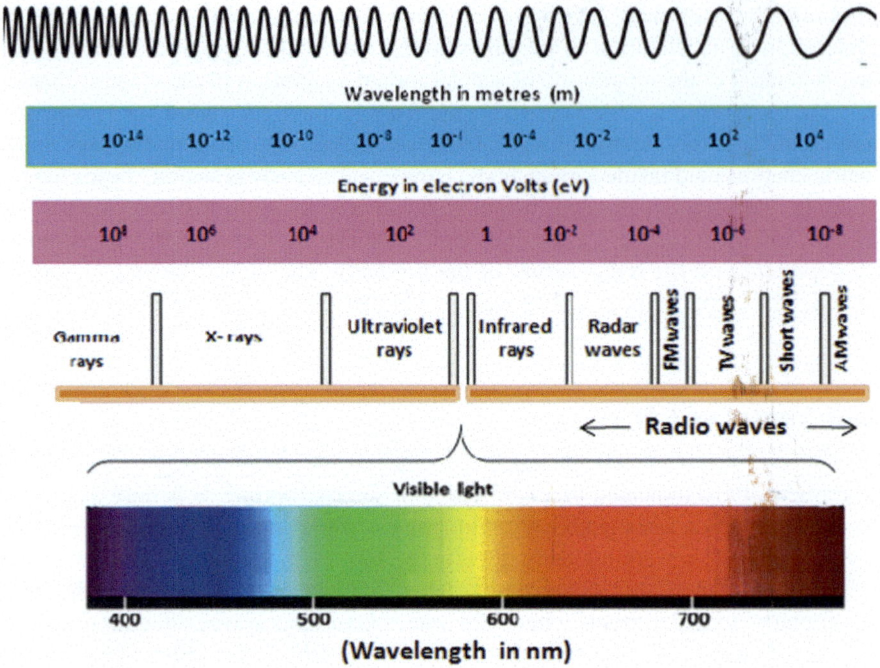

Fig. 4 Frequency/wavelength/energy spread of electromagnetic waves. Small window of visible light of different colors is also shown in the figure

Gamma rays, X-rays, ultraviolet rays, visible light, infrared rays and radio waves. These groups have overlapping boundaries and are named either on the nature of their production, their uses or their wavelength. EM waves of wavelengths lying in the small window of roughly 350–800 nm make visible light, electromagnetic waves that may be detected by the human eye. The human eye cannot detect EM waves of wavelengths smaller than 350 nm and of wavelengths larger than 800 nm. There are special instruments to detect different wavelength groups of EM waves.

Thermal Radiations

EM waves carry energy and if absorbed by an object, deposit their energy in the object as heat energy, raising the temperature of the object. It has been experimentally observed that EM radiations of smaller frequency (or of larger wavelength) are easily absorbed even in smaller thicknesses of objects, as compared to the radiations of higher frequencies. For example, EM radiations in the infrared range (wavelength 700 nm–1 mm) and of wavelengths larger than that are completely absorbed by an aluminum film of 10 μm thickness, while EM radiations of smaller wavelengths, like X-rays may pass through the same film without any appreciable absorption. Since

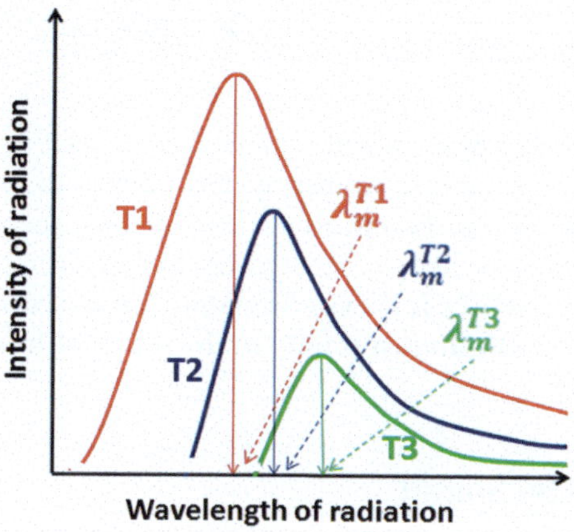

Fig. 5 Intensity distribution graphs for a body at three different temperatures T1, T2 and T3 (T1 > T2 > T3)

infrared and EM radiations of wavelengths larger than 700 nm are easily absorbed and deposit their energy in the absorbing medium as heat, raising the temperature of the medium, they are called ***thermal radiations***.

Composition of EM Waves Emitted by a Hot Body

As mentioned, a hot body emits a bundle of EM waves of different wavelengths. The intensity of each individual wavelength of the emitted group has a different value. Typical intensity distribution graphs of radiations emitted by a given body at three different temperatures T1, T2 and T3 (T1 > T2 > T3) are shown in Fig. 5. In these graphs, the wavelength and the intensity of the emitted EM waves are shown respectively on the *X*- and the Y-axes. One defines the wavelength of ***maximum emission***, denoted by $\lambda\, m$, as the wavelength that has the maximum intensity. It may be observed in Fig. 5 that $\lambda\, m$ shifts to the longer wavelength side as the temperature of the body is reduced. Experimentally, it has been found that $\lambda\, m$ and temperature T (K) obey the following relation.

$$\lambda\, m \times \mathrm{T} = \text{constant} = 2.878 \times 10^{-3}\ \text{mK} \tag{4}$$

Equation 4 tells that if the wavelength of maximum emission of the EM waves emitted by the body may be determined, it is possible to estimate the temperature T of the body.

Further, the energy **E** radiated by the given **area** of a hot body (at temperature T) **per second**, in the form of the EM waves, is found to be a function of its temperature T.

$$\mathbf{E} = \text{function of temperature T} \quad (5)$$

Equations 4 and 5 both may be used to determine the unknown temperature T of the hot body, either by determining λ_m, the wavelength of maximum emission or the energy **E** emitted per unit area per second by the hot body. Further, in the case when the body is not at a very high temperature, say, is at a temperature below 50 °C ($\simeq$ 323 K), **it is experimentally observed that the EM radiations emitted by the body lie almost totally in the range of infrared (IF) wavelengths (780–1 mm).**

The Sound

Sound waves are produced when some object vibrates or sets in vibratory motion the constituent particles of an ***elastic medium***. An elastic medium is the medium the constituent particle of which if displaced by a small amount from its original position, by applying some force, tries to regain its original position once the external disturbing force is withdrawn. The property of elasticity of a medium is the outcome of the (atomic/molecular) forces between the constituent particles of the medium. Most of the solid media (like metals, steel, copper, etc.), liquids (like water, glycerin, alcohols, etc.) and gases (like air, oxygen, etc.) are elastic media, and therefore, sound waves may be produced and propagated in these media. Unlike EM waves, sound waves need an elastic medium for production and propagation; they cannot propagate through vacuum where there is no medium (EM waves may be produced and may travel in vacuum).

Figure 6i shows how the vibratory motion of an object, like a metallic scale, may produce sound waves in the air surrounding it. The scale, clamped at the lower end if plugged, starts vibrating in the X-direction. Let us consider the vibration of the scale when it moves from its equilibrium position towards the +ve X side. As the scale moves towards the +ve X-direction from its equilibrium position, it pushes (applies a disturbing force F_D) the air particles in immediate contact with it towards the +ve X side. Since air particles some distance away are still stationary, the motion of pushed particles creates a volume of air with larger density of particles. This volume is called ***condensation***. The volume of air just behind the condensation, from where particles have moved out, becomes a volume of lower particle density. This region, where the density of air particles is lower, is called ***rarefaction***. Thus, condensations and rarefactions, one after another, generate a sound wave. Successive vibrations of the scale push this pattern of alternate condensation and rarefaction with a speed V in the given medium (air in the present case), away from the source of vibration (scale), in the +ve X-direction in the air. It may be realized that the movement or propagation of sound waves, say, in +ve X-direction essentially means the propagation of condensations and rarefactions from their respective locations further away in +ve X-direction. Similarly, the vibration of the scale in the other −ve X-direction generates sound waves consisting of condensations and

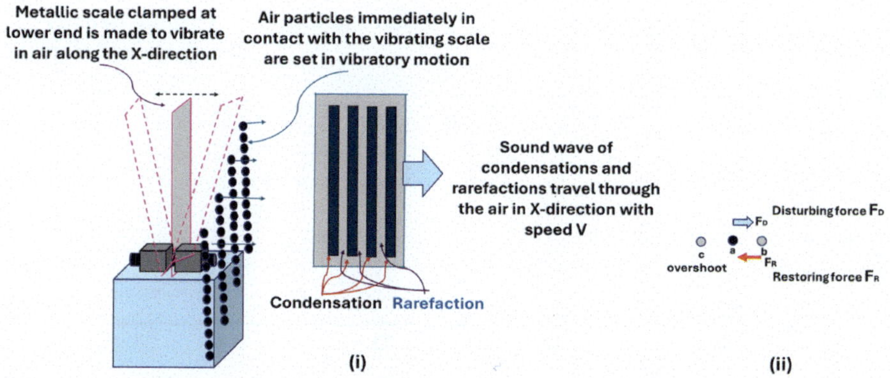

Fig. 6 **i** How the vibration of a metallic scale along the X-direction may produce a sound wave of condensations and rarefactions which moves in the air in the X-direction with speed V. **ii** The vibratory motion of an individual particle of air; the particle is pushed in the X-direction by the disturbing force exerted by the vibrating scale; however, after a small displacement up to point b the particle returns back towards its equilibrium position a due to the elastic restoring force. On reaching the equilibrium position a, the particle overshoots it and moves up to point c where it once again moves back

rarefactions that move out in −ve X-direction. Sound waves travel in different media at different speeds; for example, in the air at 20 °C the speed V of sound waves is around 344 m/s while in steel at the same temperature it is about 5200 m/s and in water at the same temperature it is 1461 m/s.

Figure 6ii shows the vibratory motion of each individual air particle; the particle under the action of the two forces; disturbing force F_D applied by the push of the vibrating scale and the restoring force F_R, that comes into play because of the elastic forces between the particles of the air, undergo vibratory motion around its equilibrium position. It may be noted that while each air particle executes a vibratory motion about its equilibrium position, the condensations (volume of higher particle density) and the rarefactions (volume of lower particle density) move through the medium (air).

It may be stressed that in the propagation of a sound wave, there is no bulk transfer of medium particles from one location to the other; each medium particle executes vibratory motion about its equilibrium position. Further, in a sound wave the direction of propagation of the wave is the same as the direction of vibrations of medium particles. Such waves are called **longitudinal** waves. The position (or location) of condensations and rarefactions in a sound wave keeps changing with time. The particle density distribution at a particular location in the sound wave, therefore, also changes with time. Suppose one takes a snapshot at some instant 't' of the particle density variation in a sound wave and plots it as a function of the distance from the source (the vibrating scale in this case). The particle density variation with distance is shown in Fig. 7 by the black curve. As may be seen, the particle density peaks at points marked by 'a' and attains the lowest value at points marked by 'b'. Small volumes around points 'a' indicate condensations (shown by red

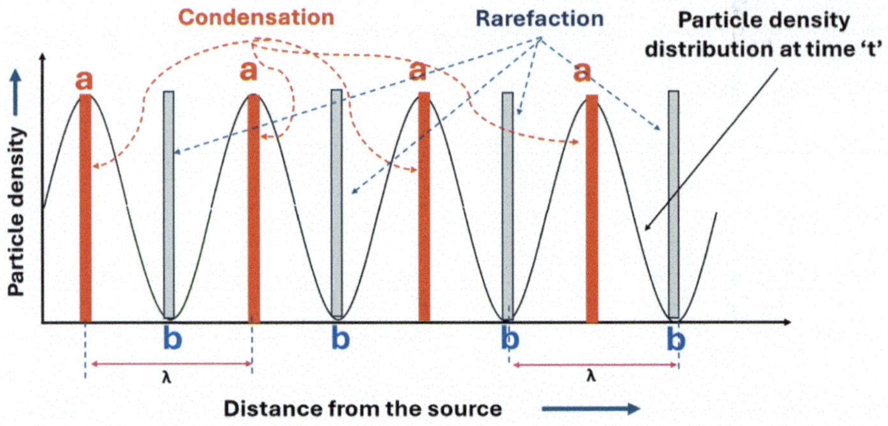

Fig. 7 Particle density variation in a sound wave as a function of distance from the source. Points marked '*a*' are regions of high density, the condensations, and points '*b*' are regions of lower density, the rarefactions. Distance between two successive condensations or rarefactions is called the wavelength λ of the wave

strips), and small volumes around points 'b' are rarefactions. The distance between two successive condensations or between two rarefactions is called the wavelength, denoted by the symbol **λ**. The frequency of vibration ***ν*** (also denoted by **f**) of the source (metallic scale in this case), the wavelength **λ** and the speed of sound **V** in the given medium (air in the present case) are related by the relation.

$$\mathbf{V} = \nu\boldsymbol{\lambda} \tag{6}$$

In the above expression, frequency ***ν*** is a parameter that depends on the source of sound while the wavelength **λ** and the speed **V** of sound are the parameters that depend on both the nature of the medium and the frequency of the source. Frequency ***ν*** or **f** is measured as the number of vibration cycles in one second (cycle/s), this unit of frequency has the name Hertz written as **Hz** in short. Speed of sound is measured in units of meter/second written as **m/s** and the wavelength **λ** in meters written in short as **m**.

Depending on the source frequency ***ν*** (or f), sound may be classified into three types: (i) ***Audio sound***, frequency **f** or ***ν*** in the range 20 **Hz** to **20 x 10^3 Hz** (20 Hz to 20 kHz), (ii) ***Subsonic sound** (also called **Infrasound**),* frequency (***ν*** or f) **less than 20 Hz** and (iii) ***Ultrasound*** frequency **above 20 kHz**. Human ears can detect only the audio sound (f in the range 20 Hz to 20 kHz). Some animals and birds have ears that may detect subsonic and ultrasonic sounds; for example, an elephant may detect subsonic sounds while some snakes and some birds (bets) navigate through ultrasound signals.

Pitch of a sound is a characteristic essentially associated with musical quality of the sound. A note of lower pitch sounds lower to our ears, while a note of higher pitch sounds higher to our ears. One may, however, classify the pitch of sound in

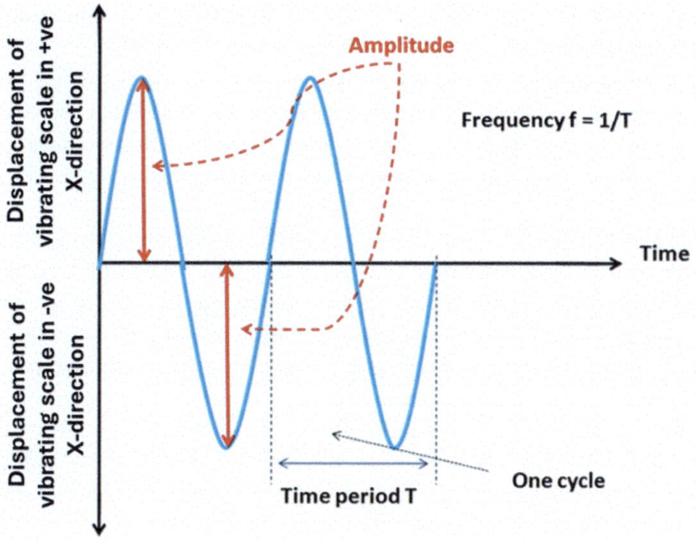

Fig. 8 Displacement of the top edge of the vibrating scale with time

terms of the frequency; a sound of higher frequency has higher pitch and a sound of lower frequency a lower pitch.

Any object or body has the tendency to vibrate if it is disturbed by hitting or by some other means. The frequency with which a body or object will vibrate on being disturbed is called the **natural frequency (ν_n)** of the object. It has been observed that the natural frequency of objects strongly depends on the size and shape of the object. Bulky or voluminous objects have lower natural frequencies, while small and lightweight objects have higher natural frequencies. Therefore, bulky and voluminous objects placed in some elastic medium, if disturbed will vibrate with lower frequency producing sound waves of lower pitch and small and lightweight objects on being disturbed will produce sound of higher pitch.

As mentioned above, in the example of the vibratory metallic scale, sound is produced by the vibratory motion of the scale in the +ve X-direction and then in the other direction. The upper tip of the scale undergoes different amounts of displacements from its equilibrium position at different times. The maximum displacement of the tip from its equilibrium position is called the **amplitude** of the vibration. The **loudness** or **intensity** of a sound wave is proportional to the square of the amplitude of the wave in a given medium. The intensity of sound is measured in a unit called **decibel** denoted by **db**. Displacement versus time graph for the tip of the scale is shown in Fig. 8.

As shown in Fig. 8, one complete cycle of vibration of the scale includes two successive vibration loops: in one direction and in opposite direction. Time taken by the vibrating scale in completing one cycle of operation is called the

time period, which is often denoted by T. The **frequency of vibration** ν (or f) is given by ν (or f) = 1/T.

An object vibrates with its natural frequency only if it is disturbed and left to vibrate. However, if the object is disturbed by a periodic force that repeats with the frequency, say after a fixed time, the object will vibrate not with its natural frequency, instead it will vibrate with the frequency of the applied periodic force. Vibrations of an object by an applied periodic force are called ***forced vibrations***.

In the case of the forced vibrations, a very interesting situation develops when the frequency of the applied periodic force ν_1 is equal to the natural frequency ν_n of the object. This is called the condition of ***resonance***. In the case of resonance, the amplitude of vibration of the object increases which in turn increases the loudness of the sound. The phenomenon of resonance is easy to observe in case of swings in playgrounds. When one pushes a swing, it starts vibrating with its natural frequency. It goes away and after reaching a maximum distance returns back. If another push is given to the swing when it comes back and this is repeated again and again, the phenomenon of resonance occurs, increasing the amplitude of the vibration of the swing.

Signals: Analog and Digital

We start by the term **Signal**; any piece of information is a signal. Signals, i.e., the (pieces of) information, may be in different forms; for example, audio sound signals are generated when we speak. They are made up of pressure (or density) variations produced in the air by the movement of our vocal cord, tongue and lips. As such sound signals are made up of air pressure variations. In a similar way, light signals consist of variations in the electric and magnetic fields of which light or electromagnetic waves are made of **Electrical Signal** which is the piece of information depicted by the variation in the magnitude of electric voltage or current, mostly voltage. To illustrate our point let us take the example of audio sound signal, as shown in Fig. 9, when some speaker speaks a sound wave, consisting of pressure (or density) variations in air is produced which moves in the positive X-direction with some velocity V.

When a sound wave consisting of pressure variations is made to hit the input of a microphone, it is converted into a wave of electrical signals as shown in Fig. 9. It may be observed that the microphone in the present example works as a device that changes the nature of the signal from pressure variations to variations in electric voltage. Devices/instruments that change the nature of the signal are given a common name ***transducer***. Transducers are available in the market that may convert a signal consisting of temperature difference or a signal of difference in the intensity of light or any other signal into corresponding electrical (or electronic) signal. Almost all medical diagnostic devices employ some sort of transducer to convert input signals of different kinds into corresponding electrical signals.

Analog and Digital Signals: The electrical signal wave (corresponding to the sound wave) shown at extreme right in Fig. 9 contains electrical signals that vary

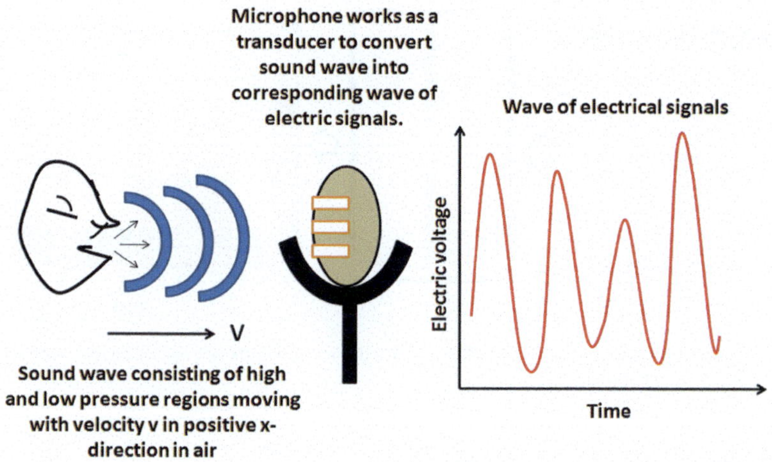

Fig. 9 Microphone works as a transducer to convert sound waves consisting of variation in pressure signals into a wave consisting of electric voltage signals

Fig. 10 A typical wave of digital signals

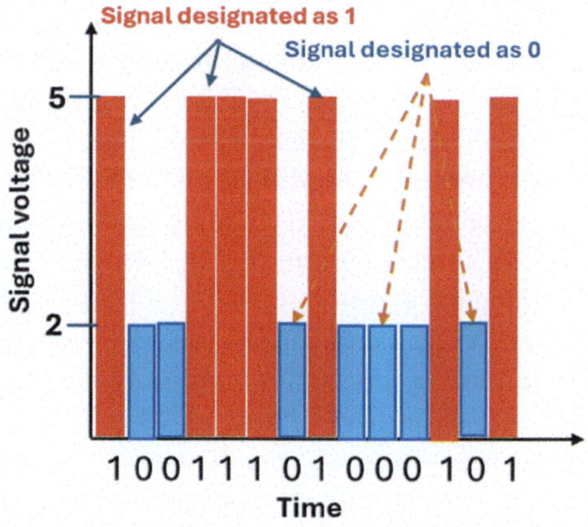

continuously in magnitude with time. Such a wave is called a wave of **analog** signals. On the other hand, a digital signal wave contains signals having only two discrete values (0 and 1) separated from each other by a fixed duration of time. It essentially means that a digital wave of electrical signals may be represented as a sequence of rectangular electronic signals with the values 1 and 0 separated from each other by fixed (but extremely small) time intervals, as shown in Fig. 10.

As shown in Fig. 10, a wave of digital electronic signals may be represented by a sequence of 1 and 0. It may be noted that a signal that is designated by 0 may not be of zero volt/current in actual case. One may define that any electronic signal less than, say, 2.0 V may be designated by digital signal 0 and a signal of, say, 5.0 V or more may be designated by digital signal of magnitude 1.

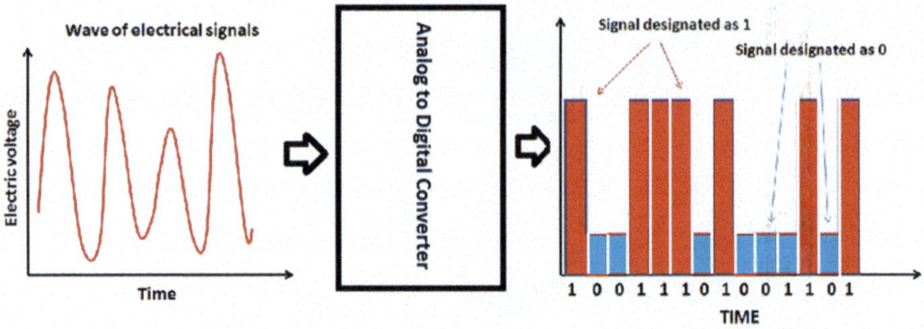

Fig. 11 Analog to digital converter (ADC or A/D) device converts a wave of analog signals into a wave of digital electronic signals

Analog to Digital (A/D) Converter

As the name suggests, an analog to digital converter device changes a wave of analog signals into the corresponding wave of digital electronic signals as shown in Fig. 11.

At present, we are not interested in operational details of the working of an ADC but would like to comment that integrated circuit (IC)-based miniature sized ADCs are now commercially available for all applications. Most of the diagnostic medical instruments/gadgets that provide digital output have ADC circuits in them.

Advantage of digital signals: Digital signals have several advantages over their analog counterparts. (i) The biggest and the most important advantage is that any piece of information in digital form is made up of only two digits; 0 and 1, while the same information in analog form may require as many as 10 different numbers, 0–9 and their combinations to specify the information. As a result, identification of the information becomes easy; only two digits need to be identified to decipher the information instead of at least 10 different digits in the case of analog signals. (ii) Very often the result of a diagnostic investigation is required to be communicated from the investigating laboratory to the concerned doctor or specialist via e-mail or Internet. This transfer of information may damage or corrupt the information during transit. The chances of damage are minimal if the information is in digital form. Digital data is less likely to get distorted and pick up system noise as the complete information is coded in only two digits 0 and 1, and the difference in the size of the corresponding electronic signals (5 and 2 V, say) is large enough to be confused. (iii) The speed of digital information transfer is much higher as compared to the speed of analog data transfer. In technical language, it is said that the ***bandwidth*** of digital transmission is large. Bandwidth in some way may be compared to the diameter of the pipe through which the information in digital form is being transmitted. The larger the diameter of the pipe, more information and at a faster rate, may be transferred. (iv) Another big advantage of digital transmission is the security of the information data; in digital communication, it is possible to encrypt the information with some sort of a key so that only those who have the key may see the data. The encrypting key is like the 'password' of your laptop or mobile.

Decimal and Binary Number Systems

Since digital communication is based on the binary number system with basic elements 0 and 1, it is required to know a little more about the decimal and binary number systems. We are conversant with the decimal system of numbers, which is a place value system of base −10. In this number system, each digit of the given number has a cipher value and a place value. Decimal numbers are often specified with a subscript (decimal); for example, decimal number $5402.12_{(decimal)}$ may be written as.

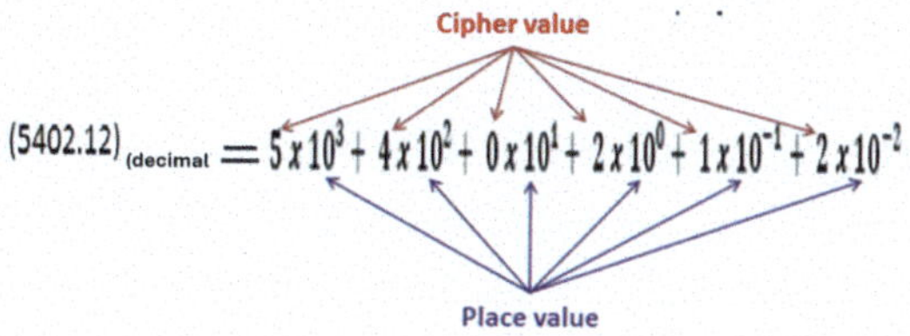

(7)

It may be observed in Eq. 7, and each digit of the decimal number has a cipher number which is a digit with some value between 0 and 9 and a place value that is some power of ten.

A number of the binary systems (specified by subscript '(binary)' also have a cipher value and a place value; however, in this system the cipher may have only two values, either 1 or 0, and the place value is expressed as some power of 2. For example, the binary number $1011.01_{(binary)}$ may be written as:

$$\begin{aligned} 1011.01 \text{ in binary equals } & 1 \times 2^3 \\ & + 0 \times 2^2 + 1 \times 2^1 + 1 \times 2^0 \\ & + 0 \times 2^{-1} + 1 \times 2^{-2} \end{aligned} \tag{8}$$

On simplifications, it gives

$$\begin{aligned} 1011.01(\text{binary}) &= [1 \times 8 + 0 \times 4 + 1 \times 2 + 1 \\ & \quad \times 1 + 0 \times (1/2) + 1 \times (1/4)](\text{decimal}) \\ &= 11.25(\text{decimal}) \end{aligned} \tag{9}$$

It may be observed in Eq. 8 that a six-digit binary number (1011.01) is equivalent to a decimal number (11.25) of only four digits. It is expected that a larger number of digits will be required to write the binary equivalent of a decimal number since the base of the binary number system is only two.

Equations 8 and 9 tell how to write the decimal equivalent of binary number, the given binary number may be looked at as the combination of two parts: an integer part on the left of the decimal and a fractional part on the right of the decimal. For example, in binary number $11100.10_{(\text{binary})}$ the integer part is 11100 and the fractional part is 01. Digits, both in the integer part and the fractional part, are the cipher values. In order to convert the binary number into corresponding decimal number, the cipher values are multiplied by the powers of 2 starting with 2^0 immediately at the left of the decimal and increasing to $2^1, 2^2, 2^3$...etc., at successive cipher values on the left to the decimal. The cipher values on the right to the decimal are multiplied by the negative powers of 2, starting with 2^{-1}. All these terms are then added to get the equivalent decimal number. For example, let us take the binary number $11100.10_{(\text{binary})}$ that may be expanded as

$$\begin{aligned} 11100.10_{\text{binary}} &= 1 \times 2^4 + 1 \times 2^3 + 1 \times 2^2 + 0 \times 2^1 \\ &\quad + 0 \times 2^0 + 1 \times 2^{-1} + 0 \times 2^{-2} \\ &= 16 + 8 + 4 + 0 + 0 + 1 \\ &\quad \times (1/2) + 0 \times (1/4) \\ &= 28.50 \text{ (decimal)} \end{aligned}$$

Hence, the given binary number is equivalent to the decimal number $28.50_{(\text{decimal})}$. It is left as a home exercise to find the decimal equivalent of binary number $0110.01_{(\text{binary})}$.

It is fun to convert a decimal number into the corresponding binary number. The integer part and the fractional parts of a decimal number are separately converted into their binary components. Let us illustrate it by taking the example of decimal number $5304.75_{(\text{decimal})}$. We start with the integer part of the number, i.e., 5304. The integer part of the number is divided by 2, and the quotient and remainders are given in separated columns as shown in Table 1 of Fig. 12. Quotient obtained after each division by 2 is further divided by two, and the remainder is stored in the last column and so on. Finally, when one gets 0 as the quotient on division by 2, further division is stopped and the reminders collected in the last column of the table are put sequentially from bottom to top to get the binary equivalent of the integer part of decimal number. The whole process of obtaining the binary number corresponding to the integer part $5304_{(\text{decimal})} = 1010010111000$ is illustrated in Table 1 of Fig. 12.

Next, we find the binary number corresponding to the fractional part of the decimal number $0.75_{(\text{decimal})}$. To obtain it, we multiply the fractional part by 2 and retain the integer value, again multiplying the fractional part by 2 till only an integer value is reached.

Integer part of decimal number/quotient divided by two (2)	Quotient	Remainder
5304 ÷ 2	2652	0
2652 ÷ 2	1326	0
1326 ÷ 2	663	0
663 ÷ 2	331	1
331 ÷ 2	165	1
165 ÷ 2	82	1
82 ÷ 2	41	0
41 ÷ 2	20	1
20 ÷ 2	10	0
10 ÷ 2	5	0
5 ÷ 2	2	1
2 ÷ 2	1	0
1 ÷ 2	0 STOP	1

Binary equivalent of decimal integer
$5304_{(decimal)} = 1010010111000_{(binary)}$

Fig. 12 Table showing the method of converting the integer part of a decimal number into equivalent binary number

Therefore, the binary equivalent of fractional part $0.75_{(decimal)}$ is $0.11_{(binary)}$. Hence, the binary equivalent of the complete decimal number $5304.75_{(decimal)} =$ 1010010111000.11.

Writing a Text in Binary Form

Since in digital representation all information is first converted in binary code, information in the form of say, English text, needs to be converted into binary form. Therefore, a binary equivalent of each alphabet of the English language has been developed. Some rules have been framed for this purpose. For example, uppercase English alphabets start with the prefix 010 while the lowercase letters start with 011. After these prefixes, one counts the number of the alphabet in the English alphabet list and writes that number in binary form. For example, the letter ‘d’ is in the lower case, and it is the fourth letter in the alphabet list. Now number four (of the decimal system) is equivalent to 00100 in the binary system; hence, the binary equivalent of letter ‘d’ is 01100100. Binary equivalents and the electric binary signals for some (decimal system) digits and English letters are given in Fig. 13.

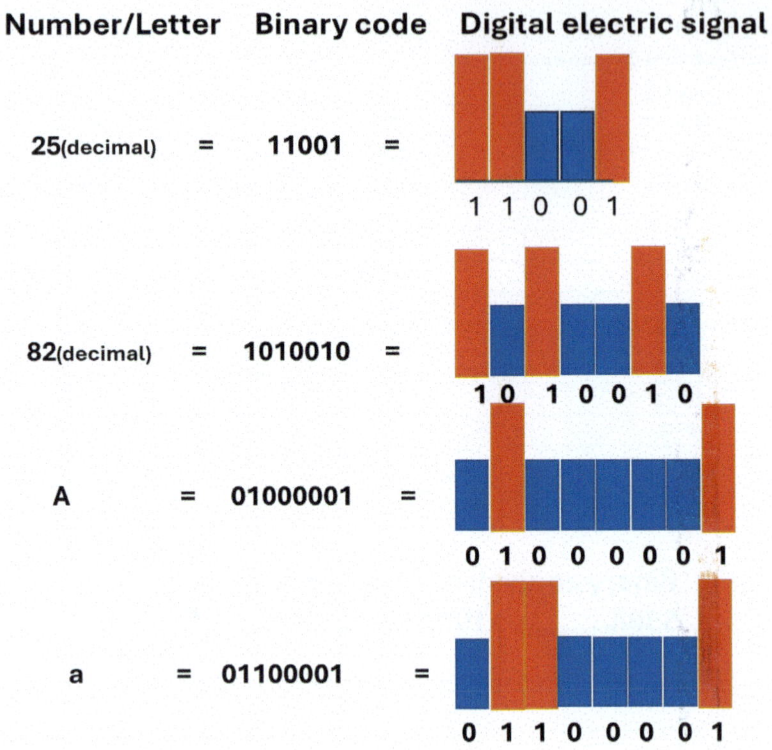

Fig. 13 Some decimal numbers and English letters with their binary codes and corresponding digital electric signals

Apart from the decimal and the binary number systems, there are also other number systems like the Octal, Hexadecimal (or Hex), Binary Coded Decimal (BCD) number and Alphanumeric codes that are in use for digital communication.

Digital Display (Signage) Screens

Most of the medical diagnostic instruments that record a particular parameter in digital form also display the measured value on some screen. Digital signage software is growing in popularity, making its mark as a popular tool for displaying, marketing and communication. Digital displays are a type of flat panel screen that uses different technologies to present digital data. The most common types of digital displays are the ***Light Emitting Diode (LED) Screens and the*** **Liquid Crystal Display (LCD) Screens**.

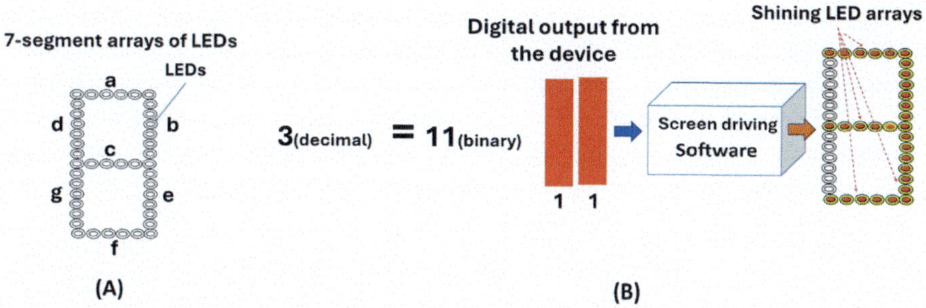

Fig. 14 **A** Seven-segment LED arrays to display digital (binary) number data in decimal number form **B** the example when the digital output from some device is $11_{(binary)}$ which is converted to 3(decimal) and the driving software apply voltages to the segments a, b, c, e and f to generate illuminated figure of three

Light Emitting Diode Display Screen Diode is an electronic device, generally made up of semiconducting materials. A special diode made using especially synthesized semiconducting materials like the Indium Gallium Nitride (InGaN), Aluminum Gallium Indium Phosphide (AlGaInP) and Aluminum Gallium Arsenide (InGaAs) emit light respectively of colors blue/green, yellow and red, when a suitable electric voltage is applied to these LEDs. LEDs are very small in size, and arrays of these diodes may be arranged in different patterns and shapes, resembling alphabets/letters and numbers of decimal system. The display panels containing different LED patterns of the alphabets/ numbers, etc., are run by a software that activates the appropriate LED array by supplying it the desired electric voltages. Thus, the human body parameter measured by a diagnostic device and represented in digital form may be displayed on the LED screen through the shining of the appropriate array of the light emitting diodes. As a typical example, the seven-segment LED arrays shown in Fig. 14A are often employed to display decimal system numbers from 0 to 10. As shown in this figure, the seven-segment figure is made up of seven arrays of LEDs (a, b, c, d, e, f) in each of which few LEDs are arranged in a straight line. These straight arrays are arranged in the form of two cubes, one above the other as shown in the figure. The output digital data from some instrument is fed to the ***screen driving software*** that issues instructions to apply appropriate voltages at different segments to illuminate them accordingly.

An example is shown in Fig. 14B. Suppose the binary output from some device is $11_{(binary)}$ which is equivalent to $3_{(decimal)}$. When this output is fed to the screen driving software, it applies appropriate voltages to the a, b, c, e and f segments which shine showing the digit 3.

It is important to note that in LED displays light is emitted by the diodes, and therefore, the screen will be visible even in a dark place.

Liquid crystal display screen LCD technology is based on the light transmission properties of some specially designed liquid crystals. In our primary classes, we are told that matter may stay essentially in three states: the solid, the liquid and the gas. In a solid, the molecules have fixed and rigid positioning with respect to each other which is difficult to change. On the other hand, in a liquid the molecules may orient themselves in any direction with respect to other molecules, and therefore, a liquid may assume the shape of the vessel in which it is placed. Liquid crystals, as the name suggests, are materials that are essentially liquid, but their molecules are not so free to change their orientation with respect to the other molecules; therefore, molecules of a liquid crystal retain their form to some extent. As such liquid crystals are like a solid, however, they are called liquid crystals because this crystalline property of liquid crystals gets completely removed by very little heating. A solid, however, requires a large amount of heat to be converted into liquid. Therefore, liquid crystals are closer to liquids than to solids.

There is one essential difference between an LED and an LCD screen. While in a LED screen light is emitted by the shining LEDs in the array, and therefore, the displayed data is visible even in a place where the background is dark, like a dark room. Contrary to that, in an LCD screen nothing will be visible if the screen is kept in a dark place. No light is emitted by the liquid crystals used in these screens. There must be background light which is reflected by a mirror included in the display screen and projects the desired figure/number or letter. That is why LCD screens show images only in a lighted place.

The working of an LCD screen is quite involved and requires knowledge of processes like the polarization of light, nematic phase liquid crystals, twisted nematic, thermotropic LCD, etc. Light, as we know, is made up of time varying electric and magnetic fields. In ordinary rays of light, the electric and magnetic fields may be confined in any plane. If an ordinary ray of light is passed through a polarizer, the light coming out of it has its electric or magnetic field confined to only one plane. Such a ray of light is called the polarized light. If a similar polarizer is put at 90° with respect to the first polarizer, the second polarizer will not allow the polarized light to pass through. This is shown in Fig. 15.

Nematic phase is the simplest liquid crystal phase. In this type of liquid crystals, the crystals are of rod shape, and they stay suspended randomly with their axis parallel to each other as shown in Fig. 16. Another important property of nematic liquid crystals is that their light transmission property, i.e., whether a polarized ray of light will be able to pass through the nematic crystal assembly, depends on the electrical voltage applied to the assembly. For example, if a layer of nematic liquid crystal assembly is sandwiched between two transparent conducting electrodes and an electric voltage, say, V_1 volt is applied to the electrodes, then it is possible that the nematic liquid crystal layer may not allow a polarized ray of light to pass through, while the layer remains fully transparent to the incident polarized light when there is no or zero voltage between the conducting electrodes.

Transparent conducting electrodes are made from materials like tin-oxide sheets that are transparent to visible light and are also conducting like a metal sheet.

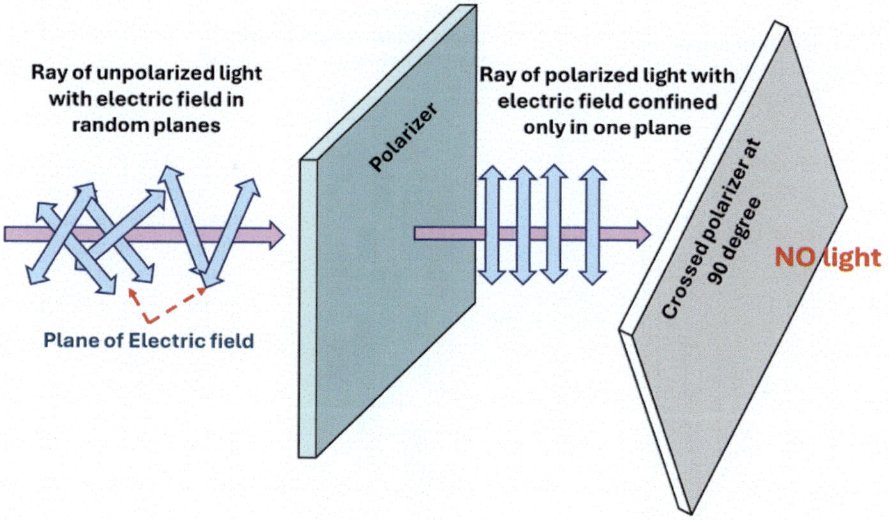

Fig. 15 Unpolarized and the polarized light

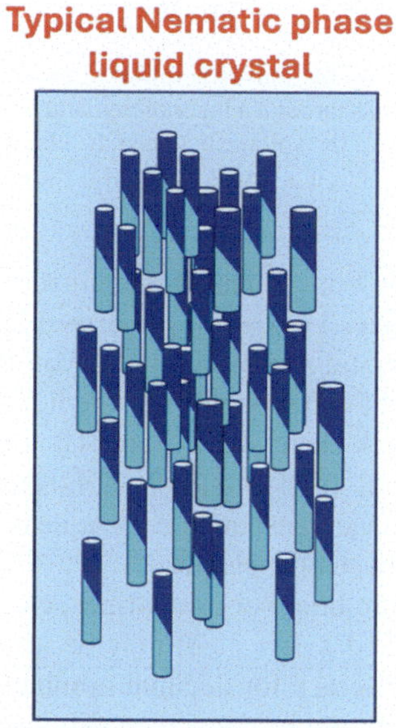

Fig. 16 Typical nematic phase liquid crystal. Rod-shaped crystals are randomly distributed with their axes parallel to each other

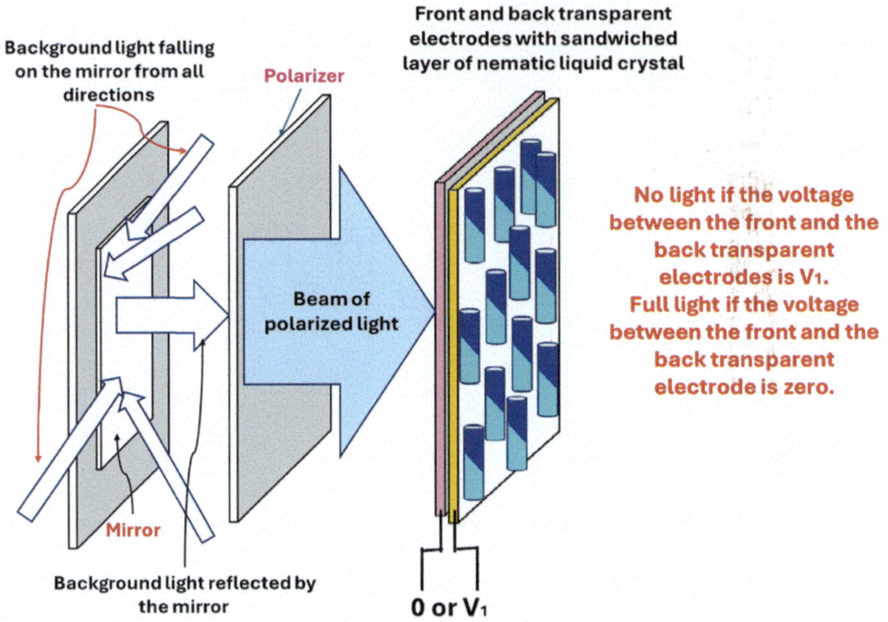

Fig. 17 Background light falling on the mirror is reflected to a polarizer that delivers a beam of polarized light which is made to pass through a layer of nematic liquid crystal held between the two transparent conducting electrodes. If the voltage between the electrodes is zero, the polarized beam of light passes through the crystal, but it is totally stopped if the electrodes are at potential V_1

The arrangement shown in Fig. 17 has a plane mirror that reflects the background light falling on it to a polarizer. A beam of polarized light is then made to fall on a layer of nematic liquid crystal sandwiched between two transparent conducting electrodes. The polarized light will pass through the liquid crystal only if the voltage between the front and the back electrodes is zero. The light will be cut off when the voltage difference between the front and the back electrodes is V_1. In this way, the liquid crystal layer may be made transparent or opaque to the incident polarized light by switching the voltage between the electrodes.

Seven independent nematic liquid crystal stripes, each sandwiched between two transparent electrodes (marked a, b, c… g) may be arranged in the pattern shown in Fig. 18 to make a display screen for decimal numbers. The binary data from the measuring instrument may be fed to the screen driving software which will apply appropriate voltages to different strips to make them transparent and/or opaque, as required, to display the data in decimal number form. In the example shown in the figure, the diagnostic instrument measured a certain parameter as 101 in binary code. The screen driving software will apply zero voltage to strips a, c, d, e and f and a voltage V_1 to the remaining strips b, and g projecting the decimal number 5 corresponding to the binary number 101.

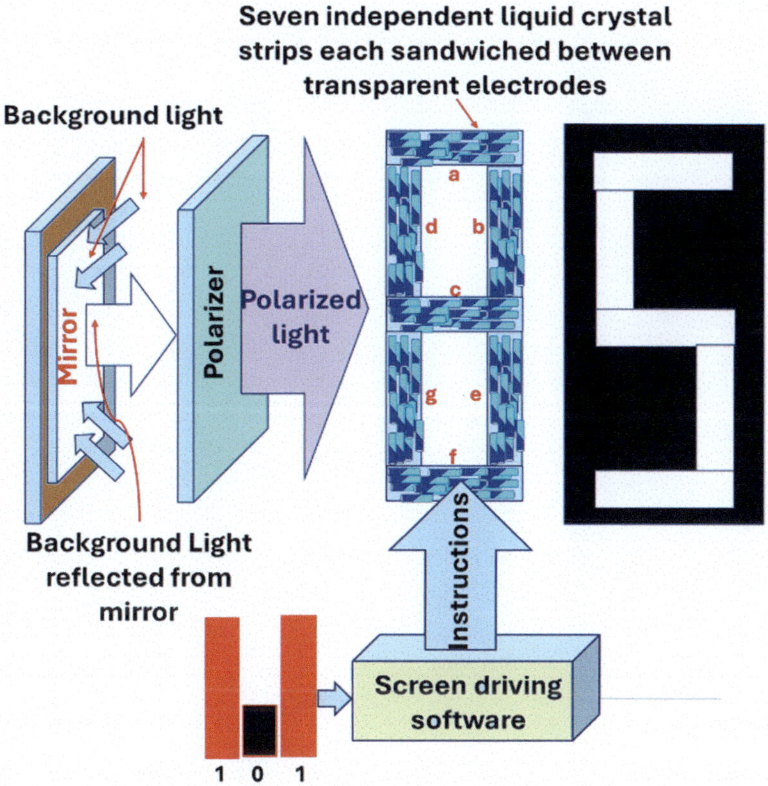

Fig. 18 Liquid crystal display system for decimal numbers

Contents

About the Author

Prof. R. Prasad is an Emeritus (retired) professor, Ex-Dean Faculty of Science and Ex-Chairman, Department of Physics, Aligarh Muslim University, Aligarh, India. He has more than forty years teaching experience of Nuclear Physics, Electronics, Thermal Physics, and other subjects to Postgraduate and Graduate students. His research specialization is Experimental Nuclear Physics, where he has supervised more than a dozen Ph.D students and has published more than 150 research papers in high impact international journals. Prof. Prasad has authored some ten books including the following.

1. Physics and Technology for Engineers
   ISBN 978-3-031-32083-5, *Springer Nature*, 2023
2. Pre-equilibrium Emission in Nuclear Reactions
   ISBN 978-0-7503-5077-8, *Institute of Physics (IOP)*, London, 2022
3. Analog and Digital Electronic Circuits
   ISBN 978-3-030-65128-2, *Springer Nature*, 2021
4. Fundamentals and Applications of Heavy Ion Collisions
   ISBN 978-1-108-49911-8 Hardback, *Cambridge University Press*, UK, 2018

5. Classical and Quantum Thermal Physics
   ISBN 1107172888 Hardback, *Cambridge University Press*, UK, 2016
6. Nuclear Physics
   ISBN 978-93-325-2265-7, *Pearson Education*, 2014

Chapter 1
The Stethoscope

Abstract Stethoscope is perhaps the most often used medical diagnostic device. In this chapter, it is explained how two basic principles of science, the principle of multiple reflection of sound and the principle of resonance of sound, are used in this device to amplify and hear the sounds produced by internal body organs.

1.1 Sensing the Internal Sounds Produced by Body Organs

Stethoscope is a device that is used primarily to listen to the sounds generated internally by heart, lungs and intestinal tract. Sounds are generated in the heart when blood flows through the heart chambers on opening and closing of cardiac (heart) valves. Flow of blood set-in vibrations in the walls of the heart producing audible sound. It may be remarked that sound waves produced by vibrating bodies may be classified into three different classes, namely those which are produced by bodies that vibrate with frequencies <20 Hz (Cycle/s), are called **infrasound or subsonic sound**. Sound produced when a body vibrates with frequency in the range 20–20,000 Hz is called ***audio sound***, as the human ears can listen or detect this sound. Sound produced by bodies that vibrate with frequency larger than 20,000 Hz is named ***ultrasound***. Both the infrasound and the ultrasound cannot be detected or heard by human ears. Further, the sound producing object in vibratory motion radiates energy in the form of sound waves. Sound waves are waves that are produced in a medium that has the property of elasticity, and therefore, sound waves are called elastic waves. Thus, sound energy travels from one place to another via sound waves, i.e., through some sort of disturbance in an elastic medium. The elastic medium through which sound waves can propagate may be a gas (like air), a liquid (like water, etc.) or a solid (like steel and other metals).

When a body vibrates, for example the scale in Fig. 1.1, it displaces the particles of the medium (air) which are in immediate contact with the vibrating object. The displaced medium particles try to regain their original position because of the property of ***elasticity*** of the medium. Thus, the vibratory motion of scale in this example sets the medium particles **(air molecules)** in vibratory motion generating a wave of

R. Prasad, *Understanding Medical Devices*, Series in BioEngineering,
https://doi.org/10.1007/978-3-031-88630-0_1

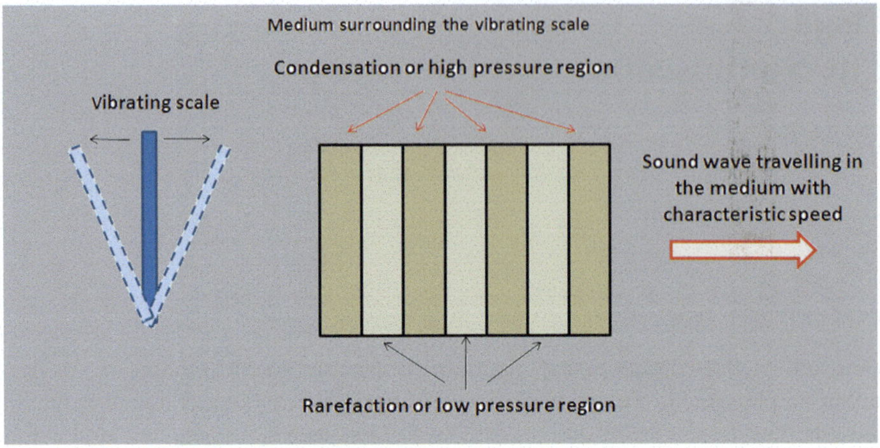

Fig. 1.1 Formation of condensations and rarefactions in the elastic medium

compression and rarefaction in the medium which travels outwards with the speed of sound in the medium as shown in Fig. 1.1. The air pressure in regions of condensation is higher than the pressure in the regions of rarefaction; a sound wave in a gas medium may therefore be called a wave of alternative high- and low-pressure regions or simply a pressure wave. When this sound wave hits the drum of a human ear, it sets the eardrum in vibratory motion of the same frequency. The eardrum vibrations, if they lie in the audio range, are picked-up by the brain which interprets them accordingly. A sound wave of high frequency is termed a high **pitch** sound while sounds produced by low frequency vibrations are termed as low pitch sound.

Natural vibrations of different body parts produce sounds of different pitches; for example, sound produced by a normal and healthy human heart is high pitch sound while the natural sound produced by the vibratory motion of healthy lungs (during inhaling and exhaling of breath) is a low pitch sound. Similarly, the sound produced by a healthy human intestine due to bowel motion is low pitch sound. These facts about the quality of sounds produced by different body parts when they are healthy and the change in the pitch of these sounds when the specific body part is not healthy were known to ancient people, like **Charaka** and **Sushruta** the two great saints who laid the foundation of medical science in India during the Vedic period (1500–600 BCE) as is evident from old sculptures of that era.

As mentioned above, sound waves in a medium are produced by the vibratory motion of medium particles which get displaced from their equilibrium or mean position, in one direction and then in the other direction. Different medium particles in a sound wave undergo different amounts of displacements from their equilibrium position at different times. The maximum displacement of a particle from its equilibrium position is called the **amplitude** of the sound wave. The **loudness** or **intensity** of a sound wave in each medium is proportional to the square of the amplitude of the wave. The intensity of sound is measured in a unit called **decibel** denoted by **db**.

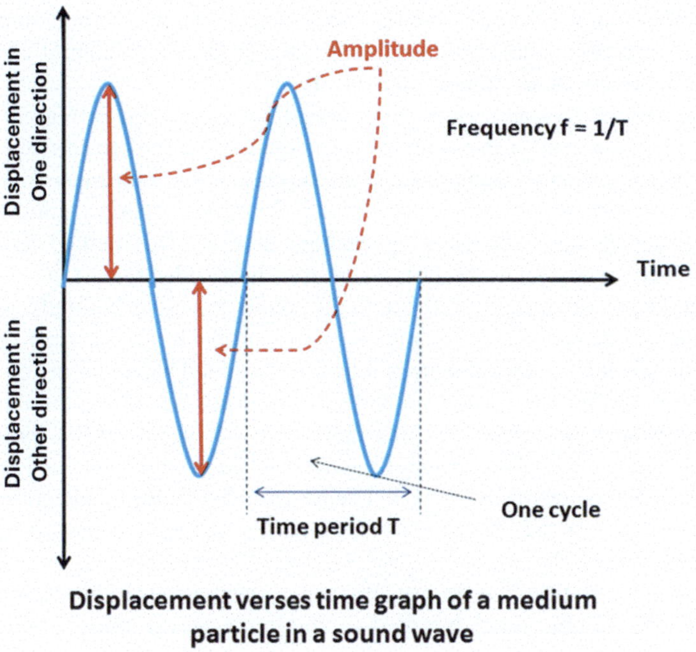

Fig. 1.2 Displacement of a medium particle in a sound wave at different times

Displacement versus time graph for a medium particle in a sound wave is shown in Fig. 1.2.

As shown in Fig. 1.2, one complete cycle of vibration of a medium particle includes two successive vibration loops: in one direction and in opposite direction. Time taken by the vibrating medium particle in completing one cycle of operation is called the **time period**, which is often denoted by T. The **frequency of vibration** f (or v) is given by $f = 1/T$. It may be realized that the intensity of the sound produced by the same source may have different values in different media as the amplitude of the sound wave produced by the same source may have different values. The parameter that remains same in all media is the frequency of vibration v, the frequency of the vibrating source.

One may, therefore, assign two different and distinct properties to a sound wave, namely the pitch that depends on the frequency of vibration of the sound producing source and secondly, the intensity which depends on the amplitude of the sound wave in a given media. Further, like any other wave, sound waves get reflected by large surfaces or obstacles.

The loudness or intensity of a sound wave of a given pitch (frequency) may be increased in three different ways: (i) by increasing the amplitude of the sound wave, (ii) by the method of multiple reflections and (iii) by the process of resonance. Method (i) of increasing the loudness of sound may be used only when it is possible to increase the amplitude of vibration of the source. Since it is not possible to do that in the case of sounds produced by internal body organs, the first method of increasing the intensity of sound is not applicable for sounds produced by natural vibration of internal body organs, heart, lungs, intestine etc.

The intensity or loudness of a given sound wave may be increased (without increasing its amplitude) if the sound wave is made to get reflected from multiple surfaces which are roughly equidistant from the observer. It is because sound waves reflected from nearly equidistant surfaces reach the observer almost simultaneously, strengthening each other, increasing the resultant intensity. This method of increasing the sound intensity is called the process of **multiple reflections**. One may easily observe the effect of multiple reflections of sound waves in his daily life; often to hear whispers or sounds of low intensity, we curl back the outer ear or earlobe inwards. The inward curved outer ear or lobe reflects sound waves falling on different parts of it towards the ear-canal leading to the eardrum enhancing the intensity of sound. In normal cases when the outer ear or lob is not curved the sound waves falling on different parts of the external ear and earlobe get reflected but do not reach the eardrum. This is shown in Fig. 1.3. The principle of multiple reflections is used in stethoscopes to enhance the loudness or intensity of feeble sounds produced by body organs.

The principle of multiple reflection of sound to increase its loudness is also used in a toy called 'Talking tube' installed in children's playgrounds. 'Talking tube' has a long metallic tube, generally of stainless-steel hidden underground with two ends protruding out of the ground at points far away from each other. The speaker whispers at one end and the receiver hears the amplified whisper sound at the other end. The whisper sound gets amplified by multiple reflections from the inner walls of the metallic tube as shown in Fig. 1.4. The figure also shows paths of two sound waves undergoing multiple reflections from the inner surface of the tube. In actual cases, many sound waves originating at the speaker end reach the ear of the observer almost simultaneously. It is true that at each reflection some energy of the sound wave is lost (by absorption) but this loss may be considerably reduced by choosing suitable material for the tube. Another example of sound amplification from multiple reflections is the production of 'echo' when a sound is produced inside a big hall;

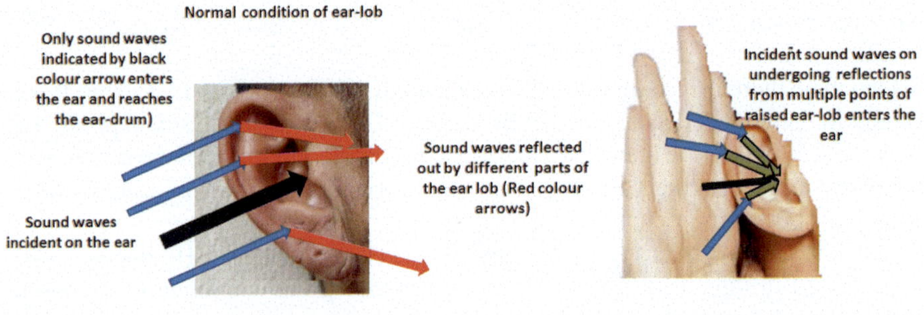

Fig. 1.3 Figure on the left shows how many sound waves incident on the ear on reflection from different parts of the earlobe do not reach the eardrum; figure on right shows how the incident sound waves on reflection from many points of the raised earlobe reach the eardrum

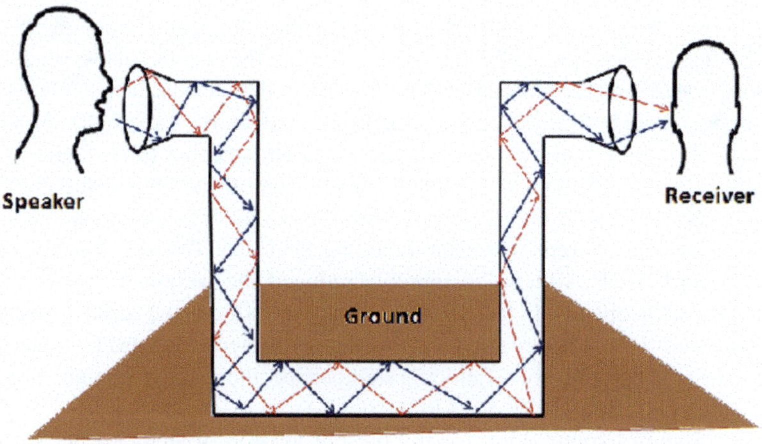

Fig. 1.4 Multiple reflections in talking tube

the sound waves reflected from the walls of the hall takes time to reach the observer (since the walls are far away) and an echo of the original sound is heard after a time delay. However, in the case of multiple reflections from a narrow tube the reflected sound waves produce a sort of echo without any time delay. Whispering galleries, found in old antique buildings, where whispers produced at one end of the gallery may be clearly heard at the other far away end, also use the principle of multiple reflections. Some famous buildings that have whispering galleries are: The Gol Gumbaz Mausoleum, Bijapur (India), Grand Central Station, New York City (USA), The Whispering Arch, Gorlitz (Germany).

A stethoscope is a device that is used by medical personnel for listening to the sounds generated by vital body organs like the heart, the lungs, the intestine of humans and animals. It may also be used to monitor the sounds produced in mechanical devices like motors, engines and other machines that have moving parts. In a stethoscope the feeble sound produced by body organs and/or moving machine parts is strengthened (or amplified) by the process of multiple reflection from the walls of the tube used in the instrument.

Most probably the crude form of stethoscope was invented by a Frenchman named Rene Laennec who worked at Necker Enfants Malades Hospital in Paris in 1816. It was a simple hollow wooden cylinder of about 25 cm length and the sole purpose of it was to avoid direct contact of the ear of the medical service provider to the body of the patient. Many old sculptures of Vedic era in India show a saint-like figure examining the abdominal and heart regions of patients with a telescope like instrument and listening to the sounds produced by internal organs. Several improvements were incorporated in the device including making the tube conical and of ivory instead of wood. However, most of the older devices required the use of only one ear of the medical observer. The precursor to the modern-day version of the instrument that uses both ears was perfected by George Philip Cammann in 1852 when he used it in his practice in New York City. In 1940, Rapport and Sprague designed a new stethoscope consisting of two sides, one of which is used to monitor the sounds (of low pitch) of the respiratory system and the other of high pitch produced by the cardiovascular system. However, the Rapport-Sprague device was heavy and short of about 60 cm and has an old-fashioned look with two large independent latex rubber tubes connecting an exposed leaf-spring joined pair of F-shaped brass binaural ear tubes with a dual faced chest piece. Several other minor improvements were incorporated in the design of the stethoscope by different scientists including Professor David Littman (Harvard Medical School, 1960) which considerably reduced the weight of the instrument and improved its acoustics. In late 1970, Littmann introduced the tunable diaphragm which increased the pitch sensing ability of the instrument. Richard DesLauriers in 1999 patented the external noise reducing stethoscope, the noise reducing component of which added considerable weight to the device.

Figure 1.5 shows some older and the present version of stethoscope frequently used by medical practitioners. A modern stethoscope may be divided into three basic components: (i) **Chest piece** which consists of a small hollow metallic conical piece at the wider end of which a fine metallic **diaphragm** (thin metallic disk that may vibrate) is fixed and on the other narrower end, called **bell**, a disk with a hole at its center is attached. The diaphragm end which is sensitive to high pitch sounds is used to pick up cardiovascular sounds while the bell side is used to pick up low pitch sounds produced by vibrations of lungs/intestines, etc. When the appropriate side of the chest piece is put above the desired organ, it picks up the sound vibrations produced by the organ and transmits them upwards in the rubber **tubing**. Multiple reflections in this tubing amplify the original feeble sound and carry it to the **headset**. The amplified sound signal from the tubing gets divided into two parts and is fed through the two ear tubes to their respective earbuds or earpieces. The medical professional stuck the earbuds in their ears. It may be noted that the process of multiple reflection also occurs in the two ear tubes. Thus, an amplified sound of a typical organ may be heard by the medical person.

To understand why the diaphragm and the bell sides of a stethoscope's chest piece are used to listen respectively to the high pitch sound of cardiovascular activities and low pitch sound of lung vibrations, it is required to understand the concepts of **natural frequency** of an object and that of ***resonance***. The natural frequency of a body (that includes objects made of solid materials, fixed amount of a gas at a given pressure and fixed volume of a given liquid at a given pressure) is the frequency with which the body will vibrate if disturbed. Each body has its own natural frequency that depends on the physical properties like the mass, shape, volume, etc., of the body. When disturbed, lightweight, small bodies may vibrate with high frequencies and, therefore, they have high value of natural frequency. Walls of the heart are light weight and have small mass, therefore, when they are disturbed by the flow of blood from one chamber to the other, they vibrate with their natural frequency which is high and generate sound of high pitch. On the other hand, organs like lungs or intestines, which have large dimensions may vibrate with natural frequencies that are low. That is the reason why sounds generated by the beating of the heart are of high pitch while those produced by lungs and intestine are of low pitch.

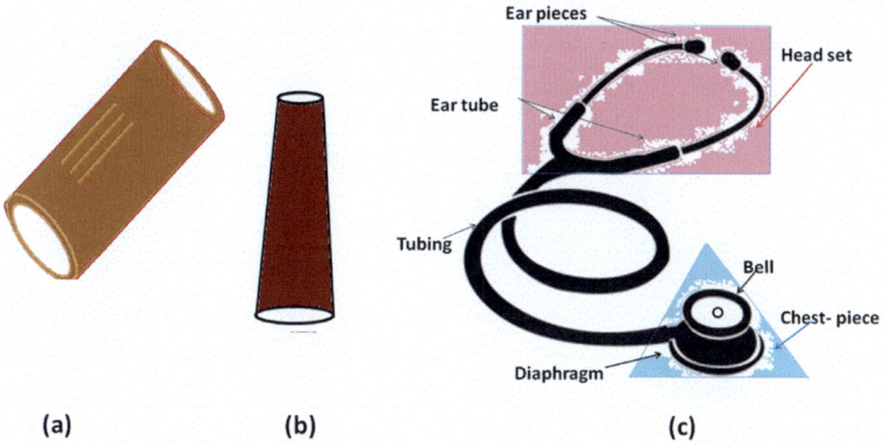

Fig. 1.5 Artists view of **a** Rene Laennec stethoscope **b** other precursors of stethoscopes **c** a modern version of stethoscope

The phenomenon of resonance occurs when a vibrating body forces another body to vibrate. If the frequency of the vibrating body is near or equal to the natural frequency of the other body, the two bodies vibrate in unison increasing the intensity of the sound. This is called resonance. The diaphragm in the chest piece of a stethoscope, generally made of either some metal or synthetic fiber, is lightweight; its natural frequency lies in the range of the natural frequency of heart vibrations. Therefore, sound wave reaching the diaphragm side of the stethoscope forces the diaphragm to vibrate and since the natural frequency of diaphragm is near to the frequency of sound generated by the vibration of heart walls, the phenomenon of resonance takes place further enhancing the loudness of sound. The bell side of the chest piece is like a hollow metallic cone, the top of which is closed by a flat plate with a hole in the center. The natural frequency of the massive metallic cone is very low. When the flat surface with a hole is put in contact with the skin above the lungs/intestine, the sound (pressure) wave pushes out excess air from the cone cavity leaving only that amount of air the natural frequency of which is same as that of the sound wave. Thus, the input low pitch sound wave vibrates in unison with the air contained in the conical cavity producing resonance and enhancing the loudness of sound. It may be noted that the hole on the flat cover of the conical bell side is required to adjust the amount of air in the cavity for resonance.

In modern digital stethoscopes, the sound received by the chest piece is fed to a tiny microphone which converts the sound pressure signal into an analog electrical signal. The electrical analog signal may be converted into the corresponding digital form using an appropriate A to D converter. The digitalized sound signal may be recorded for future use or may be transmitted to the expert/specialized doctor using the Internet. The digital signal may be converted back into analog form using a D to A converter.

Chapter 2
Thermometer

Abstract Body temperature is a crucial parameter related to the health of a human being. Mechanism of generation and control of temperature in the human body is outlined in this chapter. Principles of the working of different types of temperature measuring devices (thermometers) are also discussed in details in the chapter.

2.1 Heat Production in Human Body

Human body generates heat as a by-product of metabolic activities. Metabolic activities include chemical reactions taking place in body cells that convert sugar (carbohydrates) into ***adenosine triphosphate*** (ATP, a compound required to power other cellular processes like the flow of nutrient rich fluid across the walls of the cell, etc.), water, carbon dioxide and heat. Heat is also a form or kind of energy. Sugar in the human body comes from the food intake and is distributed to different body organs and cells through the flow of blood. In one of the most potent metabolic reactions, sugar is oxidized in body cells by the oxygen carried there by oxygenated red blood cells. The above-mentioned metabolic reaction may be written as follows:

$$\begin{aligned} \text{Sugar}(C_6H_{12}O_{16}) + \text{Oxygen}(O_2) &= \text{carbon dioxide}(CO_2) \\ &+ \text{Water}(H_2O) + \text{ATP} + \text{Heat}(\Delta H) \end{aligned} \tag{2.1}$$

Chemical reaction represented by Eq. 2.1 is the example of an ***exothermic*** reaction where heat is produced in the final state of the reaction. The rate of heat production (heat produced per unit time) depends on the speed of the exothermic reaction. Most of the heat in the human body, via the oxidation reaction represented by Eq. 2.1, is produced in the liver, brain, and heart. Some heat is also generated in skeletal muscles when one does some physical exercise.

R. Prasad, *Understanding Medical Devices*, Series in BioEngineering,
https://doi.org/10.1007/978-3-031-88630-0_2

2.2 Temperature

An object or a body that has some amount of heat energy may be associated with a parameter called temperature, often denoted by the symbol T. Temperature T of an object does not represent the total heat energy contents H of the body, though it is related to it. For example, the total heat content of a burning matchstick is much lower than the total heat content of a piece of ice, but the temperature of burning matchstick is much higher than the temperature of the given piece of ice. Temperature is an ***intensive*** property of the given body, like its density; it does not depend on the quantity of matter in it. The total heat content H of the object, on the other hand, depends on the amount of matter in the object and is an ***extensive*** property. Temperature T_A of a body A tells about the hotness or coldness of the body, more precisely, temperature T_A tells the direction of flow of heat energy when body A is thermally connected to another body B at temperature T_B. The heat energy will flow from the hotter body, the body at the higher temperature, to the colder body at lower temperature.

Temperature which is a measure of the hotness, or the coldness of a body may be measured in four different units: ***Fahrenheit*** (denoted by °F), ***Celsius*** (denoted as °C), ***Rankine*** (denoted as °R) and ***Kelvin*** (denoted as K). Fahrenheit was the first temperature scale that was proposed by the eighteen-century German Scientist Daniel Gabriel Fahrenheit and is still used; mostly in the USA, Celsius is the temperature scale commonly used in European countries and Kelvin scale, that is also called the **absolute scale**, is used to denote temperatures in scientific world. Rankine scale was initially used in engineering works, particularly on heat engines but is not in use anymore. Most of the temperature scales are calibrated with reference to the freezing and the boiling points of pure water; the freezing and the boiling points of water are taken respectively as 0 and 100 °C in Celsius scale, 32 and 212 °F in Fahrenheit scale and 273 and 373 K in Kelvin scale. The temperature of −273°C is called absolute zero, (0 K). Absolute zero temperature has not been achieved in a laboratory yet; recently the lowest temperature that has been achieved in a laboratory on earth is (0 K + 38 trillionths of a degree). Near absolute zero, some weird things start to happen. For example, it is expected that at absolute zero light will become a liquid that can literally be poured into a container. Further, NASA's ***Cold Atom Lab*** researchers claim that they have witnessed atoms existing in two places at once at very low temperatures, near absolute zero.

Following conversion factors may be used to convert temperature from one unit to the other,

$$
\begin{aligned}
&{}^\circ\mathrm{F} = 1.8\ {}^\circ\mathrm{C} + 32^\circ;\quad {}^\circ\mathrm{C} = 5/9({}^\circ\mathrm{F} - 32^\circ);\\
&\mathrm{K} = {}^\circ\ \mathrm{C} + 273 = 5/9({}^\circ\mathrm{R} - 0.6^\circ)
\end{aligned}
\tag{2.2}
$$

As already mentioned, temperature T_A and T_B of two bodies A and B tell the direction of the flow of heat energy when these bodies are brought into thermal contact. If $T_A > T_B$, then heat will flow from *A* to *B*, and when $T_B > T_A$, heat will flow from *B* to *A*.

Bodies of some living objects are found to have a nearly constant temperature, which differs from one species to another. All mammals and some birds generate heat from their metabolic activity and maintain a near constant body temperature called ***core body temperature*** and, therefore, they are called ***warm-blooded animals***. In living beings, heat is generated when food is converted into energy by metabolic action in their cells. Heat generated by cells is distributed by the blood in all parts of the body to stabilize the core temperature of the body. Healthy human body is found to have a nearly constant core temperature in the range of 97–99 °F that corresponds to 36.1–37.2 °C.

2.3 Temperature Regulation in Human Body (Thermoregulation)

Thermoregulation is the maintenance of core body temperature by balancing heat generation with heat loss. A healthy individual will have a core body temperature of 37 ± 0.5°C (98.6 ± 0.9°F); this is the temperature range needed for the body's metabolic processes to function properly. Like an electric press in which the temperature may be set to any desired value by a thermostat inbuilt with the press, the human body also has an inbuilt thermostat, called the ***hypothalamic thermoregulatory center***, which is in the ***hypothalamus***, a region of the forebrain below the thalamus, deep within the brain. This center **sets the core temperature** of the body. The hypothalamus contains temperature sensors, which receive information via nerve cells called ***thermoreceptors***. Human body has two types of thermoreceptors; the ***peripheral thermoreceptors***, located in the skin, sense skin and surface temperature and ***central thermoreceptors*** found in the viscera (soft internal organs of the body, including lungs, heart, and the organs of the digestive, excretory and reproductive systems), spinal cord and the hypothalamus, that sense the core temperature. Information received by the thermoregulatory center via thermoreceptors helps it to set the core body temperature.

Malfunctioning of the thermoregulatory center of the human body can lead to temperatures that are too low, an ailment called ***hypothermia***, or too high called ***hyperthermia***. Sometimes the set-point of the core body temperature in an individual gets raised from the normal range of 36.1–37.2 °C, which is called ***fever*** or ***pyrexia.*** This is often due to physiological processes brought about by infectious causes or noninfectious causes such as inflammation, malignancy, or autoimmune processes.

Slight temperature variations of the human body may be reversible with behavior changes and physical responses; however, extreme temperature variations may ultimately lead to organ failure, coma, and or death. It is, therefore, required to monitor core body temperature regularly. Most humans feel discomfort, body pain and uneasiness while running a fever.

A device called a thermometer is used to measure body temperature.

Most of the medical thermometers that are used to measure the temperature of human body are based on one of the following principles.

1. Expansion of liquids on heating.
2. Change in the color of liquid crystals on heating.
3. Change in the resistance of a conductor on heating.
4. Change in the voltage across a thermocouple on heating one of its two junctions.
5. Measuring the thermal energy (infrared (IR) radiation) emitted by a body.

In the following, we will discuss the working of some frequently used clinical thermometers that are based on the above-mentioned principles.

(i) **Mercury thermometer**

Most substances, solids, liquids and gases, except a few, expand on heating. In particular, metal and alloy strips, rods and liquid columns show substantial expansion in their lengths if heated. If L_0 is the length of a metal strip/rod or liquid column at temperature 0 °C and L_t at temperature t °C, then experimentally it has been found that

$$L_t = L_0(1 + \alpha t) \tag{2.3}$$

Here constant α is called the ***linear expansion coefficient*** that has a constant value for a given material. It follows from Eq. 2.3 that

$$L_t - L_0 = L_0 \times \alpha \times t \tag{2.4}$$

Since L_0 and α in the above equation are constants for a given specimen, Eq. 2.4 tells that the change in length ($L_t - L_0$) of a metallic strip/rod or liquid column due to the rise of its temperature from 0 to t °C is proportional to the temperature t. Therefore, change in the length of a given specimen may be used to determine the rise in temperature and may be calibrated in terms of the temperature. Expansion of mercury (a liquid metal) column in a capillary tube is used to make commonly used clinical thermometer shown in Fig. 2.1a.

As shown in the figure, a clinical mercury thermometer is made of a thick glass envelope which has a fine capillary at its center. The capillary is fused to a very small mercury bulb at one end and just above the bulb the capillary has a contraction. The capillary and the bulb are filled with mercury, and the other end is sealed. The mercury bulb side of the thermometer is put in contact with the body, the temperature of which is required to be measured. In the case of adults, the thermometer bulb may be put in the mouth below the tongue or in the armpit. In the case of children, the thermometer bulb may be inserted in the rectum. At room temperature or at any temperature below it, the mercury in the bulb extends only up to the contraction mark. When the temperature of the bulb increases, mercury in it expands. In an initial calibration experiment, the expansion of the mercury column in capillary is calibrated in terms of the rise in the temperature of the mercury bulb by putting the bulb at different

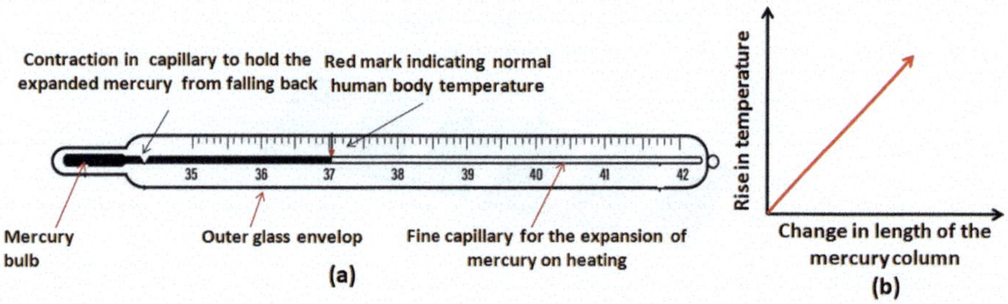

Fig. 2.1 **a** Clinical (analog) mercury thermometer, **b** Analog graph showing relationship between rise of temperature and change in length of mercury column

known temperatures. The expansion of mercury thread is thus calibrated in terms of the temperature and is marked on the body of the glass envelope. A red mark on this temperature scale on the glass envelope indicates the normal core temperature of the human body. In principle, the expended mercury column should fall back into the bulb once the source of heat (that may be the mouth, ear, rectum, armpit, etc., of a human body) is removed. However, the contraction in the capillary does not allow the mercury column above it to get back to the bulb. Thus, a permanent reading of the temperature of the source of heat is left intact even after withdrawing the thermometer bulb from the heat source. Once the temperature reading is recorded, the thermometer is given a few light jerks to bring the mercury column above the contraction back in the bulb.

Though cost-effective, simple and sturdy, the mercury thermometers are being withdrawn because of the health hazards associated with mercury.

(ii) **Liquid crystal strip thermometer**

This thermometer is in the form of a thin plastic strip that is pasted at the forehead of the patient for about a minute to record the temperature. The thermometer is made of a very thin ***liquid crystal*** which is kept in a thin plastic capsule. Liquid crystals are materials that possess the mechanical properties of a liquid but optical properties of a single crystal. It means that the structure or arrangement of molecules in the liquid crystal is not very rigid; it may get easily changed when the liquid crystal is heated. Temperature change of the liquid crystal may affect the color of the liquid crystal because of the altered crystal structure. Heat-sensitive liquid crystals that change color with the change of temperature are ca***lled thermochromic*** crystals. A forehead strip thermometer or plastic strip thermometer consists of a heat-sensitive (thermochromic) liquid crystal held in a thin transparent plastic envelope that changes color to indicate different temperatures. A strip thermometer when put on to someone's forehead changes color that depends on the temperature of the forehead. An image of a typical strip thermometer is shown in Fig. 2.2. As shown in the figure, the top surface of the strip thermometer is painted black leaving several transparent windows. Each window on the strip is marked with the corresponding temperature. When put on the forehead of a patient for about a minute, only one of

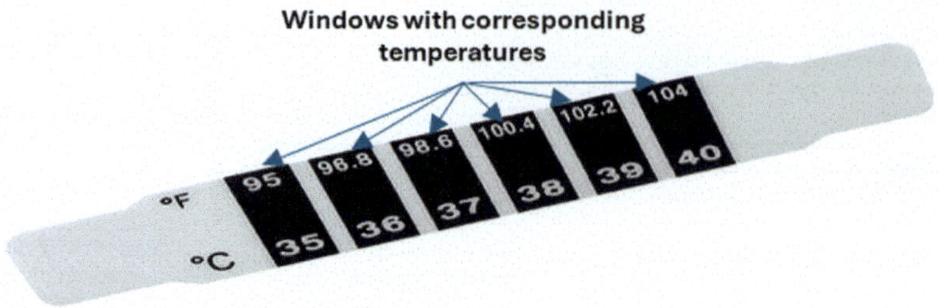

Fig. 2.2 A liquid crystal strip (forehead) thermometer

the several windows marked on the strip shines with the light of the specific color corresponding to the temperature of the forehead.

(iii) **Digital thermometers**

A variety of digital display thermometers are available in the market. All digital display thermometers have the following miniature integrated circuits (ICs)-based electronic components inbuilt within the body of the thermometer. Integrated circuit technology is the technology in which many electronic components/units, like operational amplifiers, transistors, diodes, capacitors and their interconnections required for proper processing of electric signals, are fabricated on a single miniature semiconductor chip. IC-based units require very small space and electric power to run.

(a) A transducer that converts temperature change into a suitable electrical signal.
(b) An analog to digital (A/D) converter that changes analog output from the transducer into digital form.
(c) A minicomputer with memory, which stores the calibration data for conversion of digital output from A/D converter into corresponding temperature data. The calibration curve or data may be generated by producing known temperature differences at the input of the transducer and recording the corresponding electrical signal output.
(d) A digital display screen/window generally based on liquid crystal display (LCD) technology. LCD display screens are all around us, in our laptop computers, digital clocks and digital watches, in microwaves, etc. LCD is a flat panel that uses the light modulating properties of liquid crystals. It may be noted that liquid crystals do not emit any light themselves, (in contrast to light emitting diodes (LED), LED-based display screens where diodes emit light), instead they use either a backlight or reflectors to produce multi-colored or single-color images. Some details of LCD display screens are presented in Basics.

Essentially, digital display thermometers differ from each other only in the type of the transducer used for converting temperature into electrical signals.

The working of a digital display thermometer may be explained using the block diagrams shown in Fig. 2.3a, b. As shown in these diagrams, the temperature difference *T* that is required to be measured is given at the input of the appropriate transducer. The transducer output, which is an analog electric signal, is first digitized by the A/D converter (Fig. 2.3a). The digitized output of A/D converter is then fed to the comparator/calibrator where the digitized input is compared with the digitized calibration data that is already in the memory of the computer. The digitized calibration data is stored in the memory at the time when the thermometer is manufactured, by giving temperature difference *T* of known values and recording the corresponding electrical signal values. In this way, the comparator block provides the digitized value of the temperature which is fed to the LCD for display on its screen. The other scheme of working may be the one shown in Fig. 2.3b. As may be observed, the only difference between schemes (a) and (b) is the stage at which digitization of the electric analog signal is done; in Fig. 2.3a, it is done at the initial stage while in Fig. 2.3b it is done at the end just before the LCD.

(iv) Electrical resistance-based digital thermometers

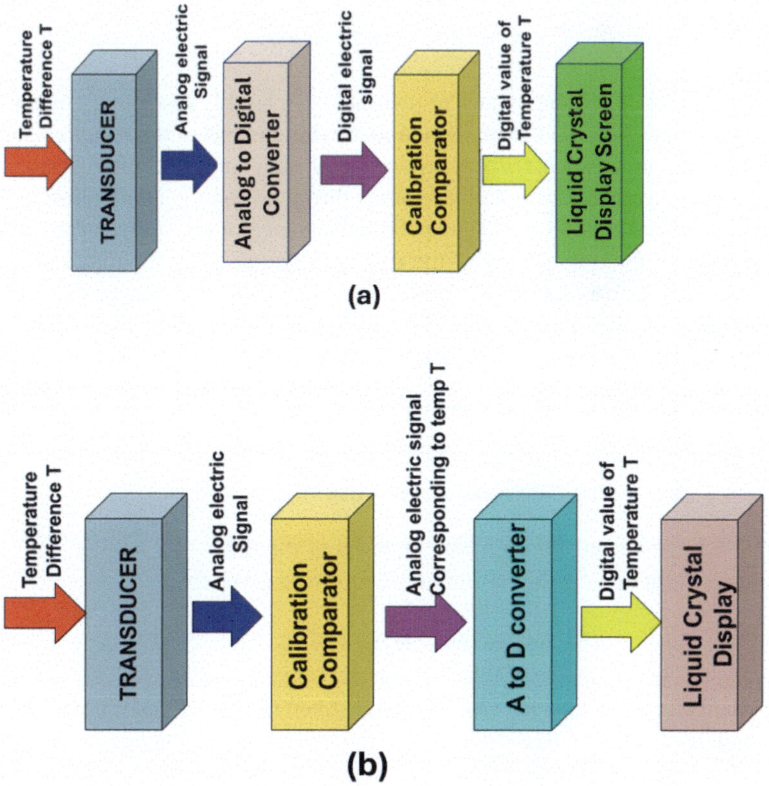

Fig. 2.3 Block diagrams to explain the working of a digital thermometer

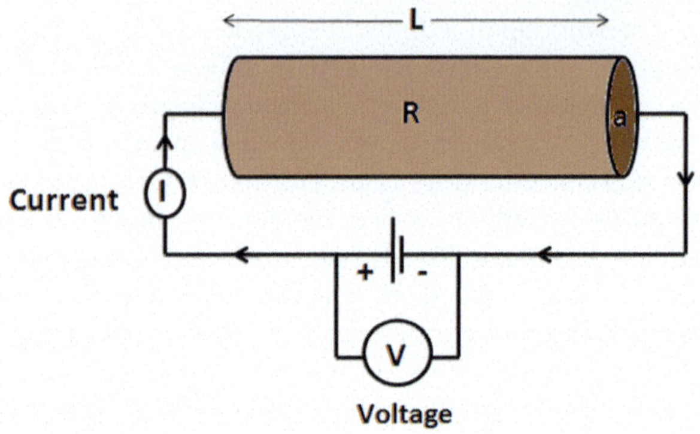

Fig. 2.4 Current flow through a conducting rod

A digital thermometer may use the electric **resistance** of a conducting rod or strip as a transducer. When a rod of length L and area of cross section 'a' of a specific conducting material, like a metal or an alloy, is connected to a source of electrical voltage V (a cell or battery) a current I pass through it as shown in Fig. 2.4. The current I and the applied voltage V are related through the relation.

$$\mathrm{V = RI;} \quad \text{and} \quad \mathrm{R} = \rho \mathrm{La} \tag{2.5}$$

In Eq. 2.5, R is called the electrical resistance (or simply resistance) of the given rod which depends both on the material and the dimensions of the rod. Resistance R is measured in a unit called ***Ohm*** represented by the symbol Ω; the resistance of a rod is 1 Ω when a current of 1 Ampere flows through it on the application of 1 Volt of potential difference across it. The specific resistance or resistivity of the material is independent of the dimensions of the rod, is the property of the material and has different values for different materials. Resistivity of a material may be defined as the resistance between the opposite faces of a cube of the material of unit dimensions.

The resistance, particularly of metal and alloy rods, increases with temperature. If R_0 and R_t are respectively the resistance of the same rod at 0 and t °C, then it is found that

$$R_t = R_0(1 + \alpha t) \tag{2.6}$$

α in Eq. 2.6 is a constant for the given material of the rod.

It follows from Eq. 2.6 that temperature t may be given by

$$t = (R_t - R_0)/(\alpha \times R_0). \tag{2.7}$$

The change in the resistance $(R_t - R_0)$ of a given rod may thus be correlated with the rise in temperature (t °C) of the body, and therefore, the change in temperature of a metallic/alloy rod may be calibrated in terms of the change in its resistance.

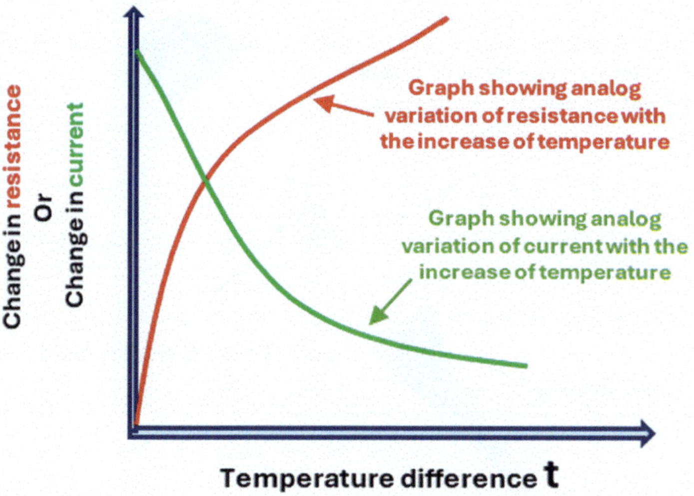

Fig. 2.5 Graph (analog) shows the variation of current and of the resistance with the rise in temperature

Referring to Fig. 2.4, when the given rod is heated its resistance will increase and the current I through the circuit will decrease, provided the potential difference V is kept constant.

Figure 2.5 shows the analog graph for the variation of current and resistance with the rise of the temperature of the rod. The analog resistance variation information, obtained using the relation ($R = V/I$), may be digitized using a miniature analog to digital converter contained within the body of the thermometer. A typical resistance-based thermometer with digital display is shown in Fig. 2.6.

The tip of the thermometer contains the resistive element. The tip of the thermometer may be put in the armpit, in the mouth, in the ear, in the rectum, etc. for recording the temperature of that specific part of the body. Stable reading is recorded in about 60 s. The temperature in digital form is displayed in the LCD window of the thermometer.

(v) **Thermocouple-based digital thermometer**

A thermocouple is a simple, robust and cost-effective temperature transducer used in a wide range of temperature measuring devices. A thermocouple is made up of wires or strips of two different metals that are joined (fused) at one end. The end of the two wires where they are fused together is called the **junction**. When the junction of the two stripes is kept at a temperature other than the two open ends of the stripes, an electrical potential difference V gets generated between the two open ends of the two wires. The induced potential difference V is proportional to the temperature difference T between the junction and the open ends of the strips. As such the generated potential difference V may be calibrated in terms of the temperature difference T between the junction and open wire ends (Fig. 2.7).

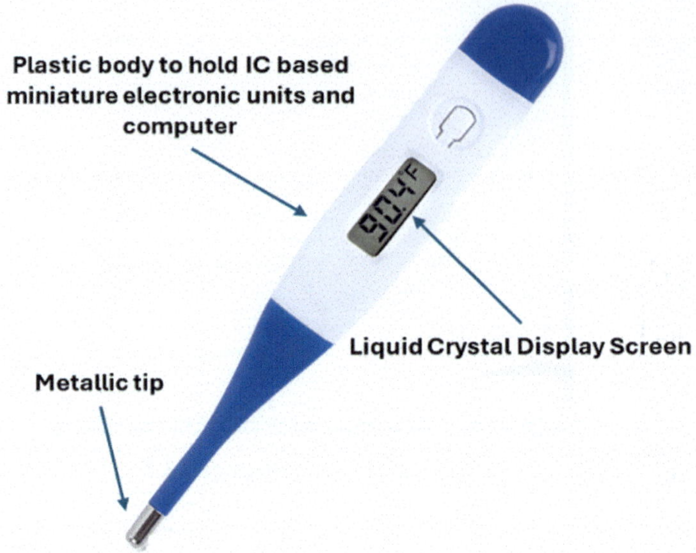

Fig. 2.6 A typical resistance-based digital thermometer

Fig. 2.7 Layout of a thermocouple

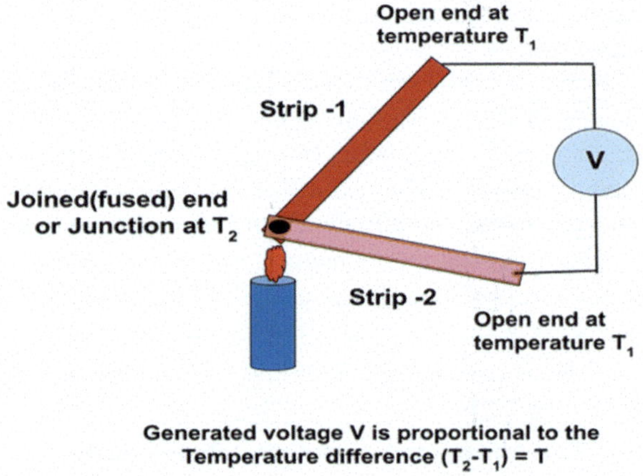

Pairs of different metal wires/strips are used in making thermocouples depending on the desired range of the temperature to be measured. Thermocouples with different pairs of metal/ alloy wires are categorized as type-J, type-K, etc. Wires of Nickel–Chromium alloy and Nickel-Aluminum alloy are used to make Type-K thermocouples that are suitable for measuring temperatures in the range of −200 to 1250 °C. A pair of a pure iron wire and a constantan (Copper–Nickel alloy) wire makes a Type-J thermocouple, which is suitable for the smaller temperature range of 0–750 °C. Since both materials (iron and constantan) are cheap, the type-J thermocouples are very cost-effective. An important parameter of a thermocouple is the output voltage V generated per degree difference of temperature of the hot and cold ends. A thermocouple-based thermometer is almost identical in appearance as a resistance-based thermometer shown in Fig. 2.6. The difference between the two is that in the case of the thermocouple thermometer, the protruding tip is the junction of the two dissimilar metal strips/wires. The two open ends of the two wires remain within the body of the thermometer and are always at the same temperature. When the tip (junction of the two wires) is kept in the mouth or the armpit of the patient, it attains the temperature of the human organ (mouth or armpit) where it is placed, setting a temperature difference between the junction and the open ends of the two wires, generating a voltage difference between the two open ends of the wires. This voltage difference is digitized by the miniature A–D converter included in the body of the thermometer. The digital value of the voltage is converted into temperature using the calibration data already stored in the memory of the instrument and is displayed on a liquid crystal display window.

(vi) **Remote sensing infrared (IR) thermometer**

Sometimes it is required to record the temperature of the patient without coming in contact with him. This happened during the COVID epidemic, it being highly contagious and may spread from one person to the other via physical contact. Infrared (IR) radiation-based forehead-sensing gun-type thermometers are used for remote temperature recording. Apart from monitoring patient temperature from a distance, infrared thermometers are used for a variety of temperature monitoring functions, for example in checking for hot spots in firefighting, in electric and mechanical equipment and remotely monitoring the temperature of an oven or boiler, etc.

As mentioned earlier, a body at some absolute temperature T K (T K > 0 K) radiates energy in the form of EM waves (it is assumed that only bodies that are at absolute zero temperature (0 K = −273 °C) do not radiate any energy). Further, to a large extent, the energy emitted in the form of EM waves by a hot body, does not depend on the nature (material) of the hot body and depends only on the temperature of the body. It means that if one takes two or more identical objects made up of different materials (say of gold, silver, brass, etc.) and heat them to the same absolute temperature T K, the electromagnetic radiations emitted per unit time per unit area of all the different objects will be identical both in nature and amount. Further, if the temperature T is not very high, most of the emitted radiation will be infrared or thermal radiation (also called heat radiation).

Thermal radiation emitted by a hot body may travel in air up to a few meters; however, their intensity decreases rapidly with the distance. This thermal radiation may be picked up by a suitable transducer placed at some distance, say D (around 1 m), from the hot body. The total thermal energy absorbed by the transducer surface in a small time interval Δt (1 or 2 s) may be used to generate an electrical signal, the intensity (or the height) of which will be a function of the temperature T of the hot body placed away at a distance D from the transducer as shown in Fig. 2.8

As indicated in the block diagram of Fig. 2.8, the remote sensing thermometer (shown by dark black rectangle) is held at a fixed distance D from the hot body (the forehead of the patient), temperature of which is required to be measured. A circular

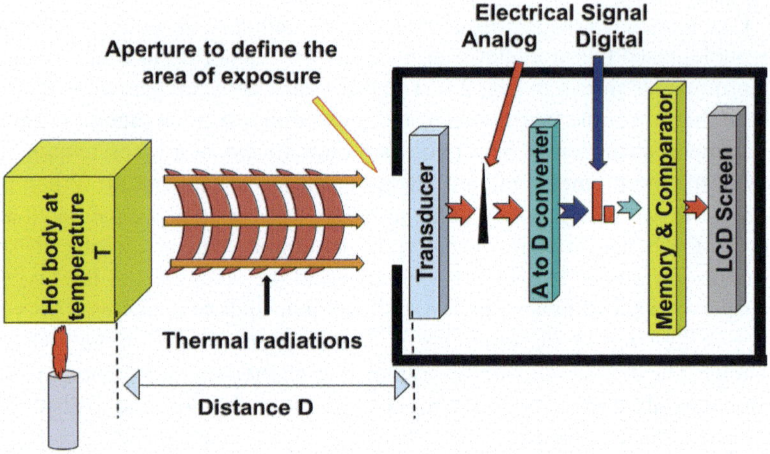

Fig. 2.8 Block diagram of a remote temperature sensing thermometer

aperture, around 2.5 cm diameter, is made in the front wall of the thermometer to fix the area of exposure of thermal radiation coming from the hot body and falling on the surface of the transducer. After a couple of seconds from the start of exposure, during which enough thermal energy is absorbed by the transducer, the transducer produces an analog electric signal in its output which is fed to an A/D converter. The A–D converter delivers a digital signal corresponding to the analog signal at input. The digital output of the A/D converter may be compared with the calibration data already stored in the memory in an earlier calibration experiment. Comparison of the digital signal with the calibration data provides the temperature of the hot body in digital form which is displaced on the LCD screen.

The calibration experiment: In an initial experiment, called the calibration experiment and carried out by the manufacturing company, the magnitude of the electrical signal from the output of the transducer (placed at a fixed distance D from the hot body and exposed to for a fixed time 'Δt' (of few tens of seconds) is calibrated in terms of the temperature of the body. This is done by heating a typical metallic ball to different temperatures and recording the strength of the electrical signal at the output of the transducer. By giving the metal ball different fixed temperatures in steps and recording the strength of the corresponding electrical signal, a calibration table or graph is generated which is recorded in the memory of the minicomputer of the device. When the temperature of some unknown hot body is required to be determined remotely, the transducer-containing device, the remote thermometer, may be kept at the same distance D from the body and exposed to thermal radiations coming from the body for the same time 'Δt'. The strength of the electrical signal (Δv) generated at the output of the transducer may be measured. The temperature of the hot body corresponding to the electrical signal (Δv) may be obtained from the calibration data, stored in computer memory that correlates the electric signal strength (Δv) with the temperature T.

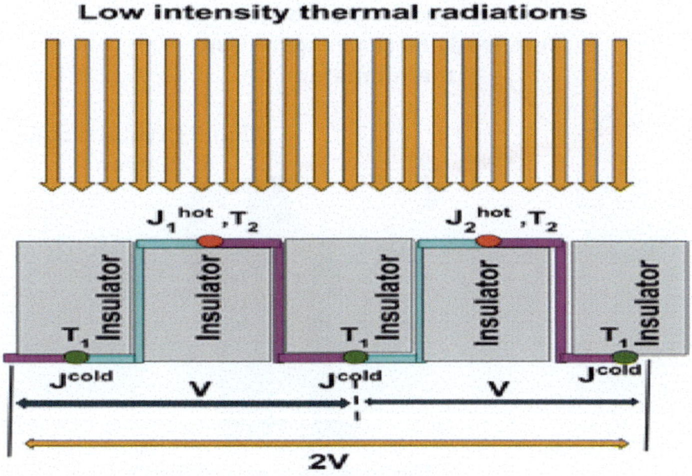

Fig. 2.9 A thermopile consisting of series combination of two thermocouples

The transducer: In a remote sensing thermometer kept at some distance from the forehead of the patient, low intensity thermal/infrared radiation reaches the transducer. Therefore, a gun-type remote sensing thermometer requires a transducer that may convert the low intensity infrared/heat radiation into electric signals. Since the intensity of the incident radiation is very low, it is required to use a transducer that may generate electrical signals of sufficient strength even with such low intensity thermal radiation. **Thermopile** is one such transducer which is simple and cost-effective. A thermopile is a combination of few (2–10) thermocouples in series. A thermopile having a series combination of two thermocouples is shown in Fig. 2.9. The hot junctions, J_1^{hot} and J_2^{hot} of the two thermocouples, exposed to the low intensity infrared (thermal) radiations coming from the forehead of the patient and falling on them simultaneously, get heated to temperature T_2.

The other junctions, called cold junctions and designated as J^{cold} in the figure, remain at the lower temperature T_1 that is the temperature inside the body of the thermometer. Since there is insulating material between the arms of the thermocouples, the heat carried by the thermal radiation does not reach the cold junctions. Each thermocouple of the series combination produces a small voltage V that depends on the temperature difference T $(= T_2 - T_1)$. However, the net voltage at the output of the thermopile will be 2 V as shown in the figure. In this way, if there are n-number of thermocouples in a thermopile, then the thermopile will deliver a net voltage of the value nV, where V is the output of each thermocouple. Since thermopiles are simple and cheap, they are used as a transducer in most of the gun-type remote sensing clinical thermometers.

A typical remote sensing gun-type thermometer is shown in Fig. 2.10. While measuring the temperature, the front end of the gun is kept at the level of the forehead of the patient, pointing directly towards the center of the forehead. It is important to

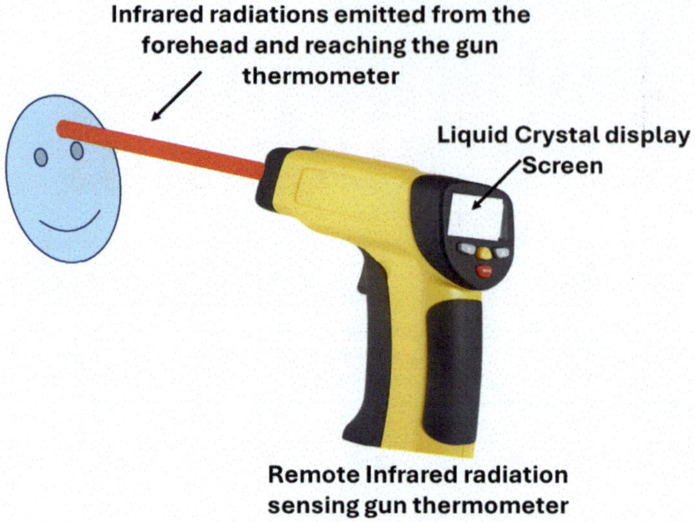

Fig. 2.10 A typical remote sensing gun-type thermometer

remember that while using this thermometer, the distance D between the forehead of the patient and the front of the thermometer gun should be kept as specified by the manufacturer of the device for a reliable reading.

Chapter 3
The Pulse Oximeter

Abstract Circulating oxygenated blood to different body parts and organs is vital for the energy generation and survival of the human body. Blood pumped by heart circulates through the fine capillaries in lungs where the red blood cells absorb oxygen from the air inhaled. In a healthy person, it is expected that 95–100% red blood cells get saturated with oxygen when passing through the lungs. However, some ailments reduce the oxygen saturation level of blood, and it is therefore required to monitor the oxygen saturation level of blood frequently in such cases. Since the oxygen intake by red blood cells also depends on the breathing rate, it is required to monitor the breathing rate also. Pulse oximeter serves the dual purpose of monitoring the oxygen saturation and the breathing rate in a patient. The instrument works on the principle that the oxygenated and the unsaturated red blood cells have different values for the absorption of the red-light and the infrared light. Details of the method along with the constituents of the human blood and its grouping are discussed in the chapter.

3.1 Introduction

In this and following chapters, we shall be talking about oxygen level in blood, blood pressure, pulse rate, etc.; some parameters associated with the blood and its flow to different human body organs. To appreciate the significance of these parameters, it is required to know, in some detail, as to what blood is, where and how it is produced and circulated in the human body.

3.2 The Blood

Blood is a vitally important fluid for the human body. It is reddish in color, denser than water and feels a bit sticky. The average temperature of blood in the human body is 38 °C (100.4 °F) which is about 1 °C higher than the average body temperature.

R. Prasad, *Understanding Medical Devices*, Series in BioEngineering,
https://doi.org/10.1007/978-3-031-88630-0_3

3.3 Important Components of Blood

There are four important components of blood: (a) ***plasma,*** (b) ***red blood cells*** (also called Erythrocytes, short form RBCs), (c) ***white blood cells*** (Leukocytes) and (d) ***platelets*** (cells).

Blood is made up of about 55% plasma and about 45% of three types of blood cells.

3.4 Plasma

The liquid component of blood is called **plasma,** and it is pale or dull reddish in color. It is a mixture of **water** (about 91%) and around 9% **solids** that include **coagulants** (a type of protein that helps in stopping the loss of blood from a ruptured blood vessel), mainly ***fibrinogen***; **sugar**, **plasma proteins** such as ***albumin*** and ***globulin***; **electrolytes** like ***sodium, potassium, bicarbonates, chloride, and calcium***; **immunoglobulins** that help in fighting infections; and various other small amounts of **enzymes** and ***vitamins***. The origin of plasma is interesting because no specific body organ produces it. Plasma is formed from water and salts absorbed through the digestive tract.

3.5 Red Blood Cells (RBCs)

Red blood cells are bright red in color and are the most abundant cells in blood, accounting for about 40–45% of its volume. In shape, RBCs look like a donut without a hole at the center, instead a depression, like a biconcave disk with a flattened center. The size of RBC varies between 6 and 8 μm (micrometer). The average lifespan of RBCs is around 120 days (three months). RBCs are produced in ***bone marrow***. Unlike many other cells, RBCs have no nucleus and, therefore, are flexible; they may easily change their shape, helping them fit through various blood vessels in the body. Because of the frequent change in shape while passing through blood vessels, the membrane of red blood cells gets damaged, and they lose their energy contents ultimately ending their lifespan in about 120 days. Red blood cells contain a special protein called ***hemoglobin***, which is red in color. RBCs get their red color because of the hemoglobin. Hemoglobin helps carry oxygen from the lungs to the heart that distributes it to all body tissues and organs. It also returns carbon dioxide from different body parts to lungs so it can be exhaled.

3.6 White Blood Cells (WBCs)

WBCs are in a very small number in blood, only about 1% of the blood. The white blood cells are larger in size than the RBCs; they may be 12–17 μm in size, round with a distinct center membrane or nucleus. Their lifespan ranges from 12 to 20 days. White blood cells are actually not white, and they are colorless. WBCs protect the body from infection. There are two important types of white blood cells, the ***neutrophil*** and the ***lymphocyte***. Of the total WBCs roughly 55–70% are neutrophils. Each neutrophil lives less than a day, and therefore, the bone marrow must constantly produce new neutrophils to maintain protection against infection. The other class of WBCs, lymphocytes, are of two types: the ***T-lymphocytes*** that help regulate the function of other immune cells and directly attack various infected cells and tumors. The ***B-lymphocytes*** make antibodies. Antibodies are proteins that specifically target bacteria, viruses and other unwanted foreign material.

Platelets (thrombocytes).

Platelets, unlike RBCs and WBCs, are not cells but small fragments of cells. Platelets help in blood coagulation or clotting when a blood vessel gets broken. Platelets gather at the site of an injury, sticking to the lining of the injured vessel, forming a platform on which blood coagulation may take place. As a result, a clot of fibrin is formed that covers the injury and prevents blood from oozing out. New tissues are then formed on the fibrin clot promoting healing of the injury.

The average number density of platelets in a healthy person may vary from 150,000 to 450,000 platelets per microliter of blood. Number density of platelets lower than 150,000, called ***thrombocytopenia***, may lead to excessive bleeding. On the other hand, platelet count of more than 450,000 per microliter, called ***thrombocytosis***, may lead to blood clot formation that may be a cause of heart attack. Figure 3.1a, b shows the drawings respectively of blood components and the structure of the red blood cell.

3.7 Blood Groups

Human blood is classified into different blood groups based on antibodies and antigens present in the blood. ***Antibodies*** are typical protein molecules found in the **plasma** of the blood, while ***antigens*** are proteins (protein molecules) found on the surface of the **red blood** cells. Three main types of antigens found attached to the surface of human red blood cells are named as: ***A-antigen, B-antigen and Rh-antigen***. The two main antibodies found in the plasma of human blood are named as: Anti-A and Anti-B antibodies. The blood group is designated Rh^+ (Rh positive) when the molecule of antigen-Rh is present on the surface of the red blood cells and as Rh^- (Rh negative) when it is not present. Therefore, on the basis whether antigen Rh is present or absent from the surface of the red blood cells, blood may be grouped

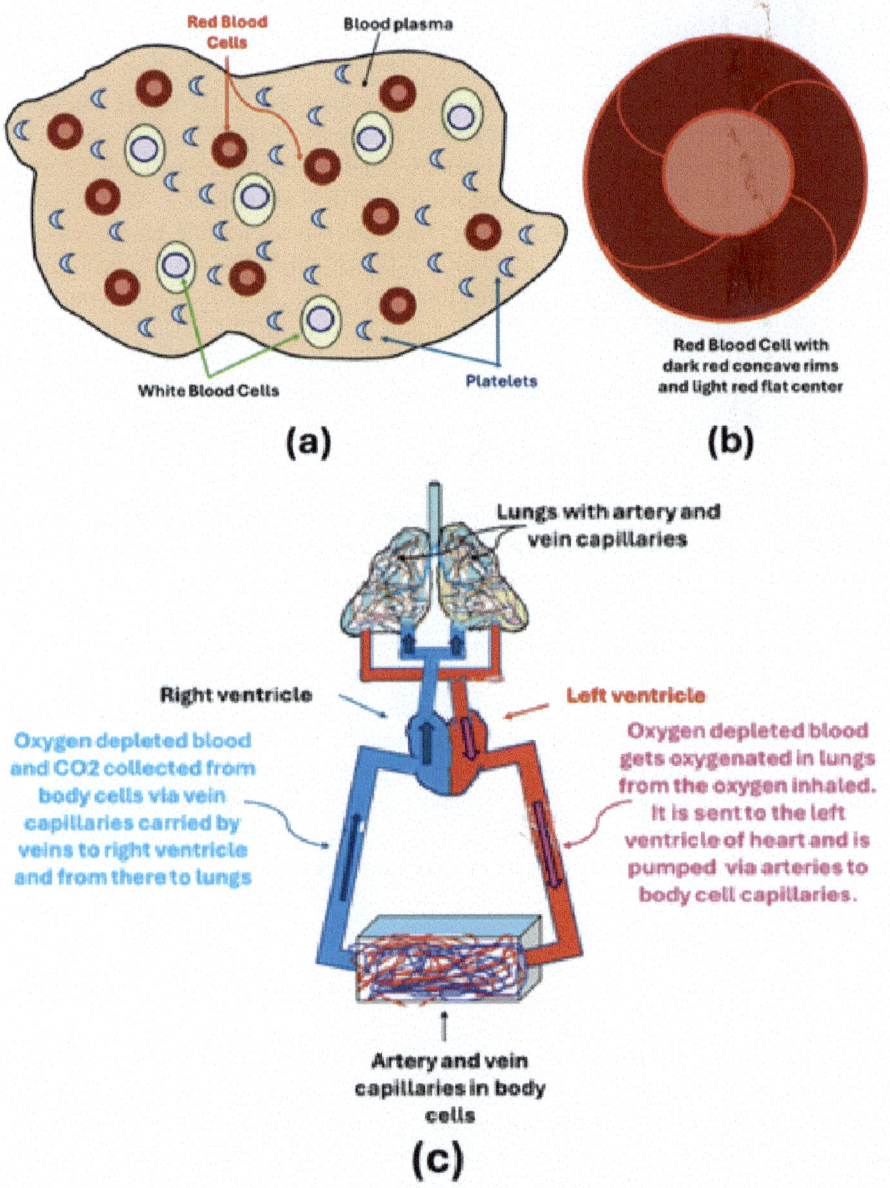

Fig. 3.1 **a** Blood and its components. **b** Shape of red blood cell. **c** Schematic diagram of blood flow in the human body

into two types: as ***Rh-positive*** (Rh^+) or ***Rh-negative*** (Rh^-). Both, Rh(+) and Rh(−) blood groups, may be further classified as:

Type A: Has antigen *A* on red blood cells and antibodies-*B* in plasma

Type B: Has antigen *B* on red blood cells and antibodies-*A* in plasma.

Type O: Has no antigen on red blood cells but both antibody-*A* and antibody-*B* molecules in plasma

Type AB: Has both antigen-*A* and antigen-*B* molecules on the surface of red blood cells but no antibodies in plasma.

In all, therefore, there are eight major blood groups written as: (i) $A(+)$ or A^+ which is *A* type blood with Rh antigen present. (ii) $A(-)$ or A^-, *A* type blood with no Rh antigen (iii) $B(+)$ or B^+ which is *B*-type blood with Rh antigen present (iv) $B(-)$ or B^- which is *B*-type blood with no Rh antigen, (v) $O(+)$ or O^+ (vi) $O(-)$ or O^- (vii) $AB(+)$ or AB^+ and (viii) $AB(-)$ or AB^-.

Blood group matching is required when a patient is given blood from an external source/donor. Serious complications occur when blood of a different group is given to a patient.

3.8 Blood Circulation in Human Body

The **cardiovascular** or circulatory system in the human body pumps blood from the heart to the lungs to get oxygen. Oxygenated blood from lungs comes back to the heart which then pumps the oxygenated blood to the brain and other parts/organs of the body through arteries and capillaries. The ***aorta*** is the main artery that carries oxygenated blood from the left side (left ventricle) of the heart to the rest of the body through many smaller arteries and capillaries. To circulate oxygenated blood to each and every organ, tissue and cell of the body, big arteries divide into smaller arteries which finally end into groups of very fine capillaries. Capillaries are very thin and may penetrate the smallest tissues and cells. Further, the walls of capillaries are such that they may deliver oxygen, other nutrients, vitamins, and hormones to the cell and/or the tissue. All body parts including cells and tissues have the distribution of two types of capillaries: capillaries that are the end parts of arteries and supply oxygenated blood, nutrients, and hormones to different organs, tissues and cells. The other groups of capillaries are connected to veins which collect oxygen depleted blood, and other wastes like carbon dioxide, from different body parts and deliver to the right ventricle of the heart. The oxygen depleted blood and carbon dioxide, collected through capillaries and veins are pumped from the right ventricle to lungs. Oxygen depleted blood in lungs gets oxygenated from the oxygen inhaled in breathing and the waste carbon dioxide is exhaled. All those parts of the body through which blood flows are collectively called ***blood vessels***, for example, arteries, veins, capillaries, tissues and cells are all blood vessels. Figure 3.1c shows the schematic diagram of blood flow in the human body.

3.9 Measurement of Oxygen Level in Blood

It goes without saying that oxygen plays a crucial role in keeping the human body alive and healthy. The body consumes oxygen essentially through ***aerobic metabolism***. Aerobic metabolism in the human body refers to chemical processes in which oxygen is used to convert carbohydrates (or sugar) in the body to energy. These chemical reactions are also called ***aerobic respiration, cell respiration or oxidative metabolism***. Oxygen is carried to the lungs with inhaled air. Lungs are made up of many fine capillaries through which blood pumped by the heart circulates. Many capillaries increase the effective total surface area of lung's capillaries that get exposed to the oxygen containing air inhaled. The hemoglobin (red blood corpuscles) present in the blood circulating in the lung capillaries, readily combine with oxygen molecules carried by the inhaled air. Each hemoglobin molecule (represented by the symbol **Hb**) has four sites where oxygen molecules may get attached and in a healthy human body almost all Hb molecules present in the blood circulating in lung capillaries get fully saturated with oxygen, i.e., each Hb molecule attaching four O_2 molecules. Oxygen-saturated hemoglobin is represented by $\mathbf{HbO_2}$ and without oxygen saturation as Hb. Each gram of 100% saturated hemoglobin (HbO_2) contains around 1.34 milli-liter (mL) of oxygen. The oxygen-saturated hemoglobin blood in the lungs is transported to the left ventricle of the heart from where it is pumped to various parts of the body through **arteries** and is ultimately distributed to the different body organs and tissues. Body tissue consists of cells which are filled with fluids that have carbohydrate (or sugar) content. Carbohydrates in the human body are produced by the process of digestion of food in the stomach and are distributed to tissue cells. When oxygen-saturated blood (pumped from the lungs via arteries) encounters carbohydrate fluid of the cell, the oxygen molecules attached to it ***burns*** (a chemical reaction in which heat is produced) the carbohydrate content-producing energy and carbon dioxide (CO_2) gas. Burning of carbohydrates results in the production of energy, necessary for all human activities. However, in the carbohydrate-burning process blood hemoglobin gets depleted of oxygen. The oxygen depleted blood along with other waste materials like CO_2 left in tissue cells is carried back to the heart through **veins** which again pump it to lungs for oxygen saturation. The by-product of carbohydrate burning, the carbon dioxide gas, is exhaled in breathing. Thus, a cycle of pumping of oxygen depleted blood to lungs, oxygen saturation of blood in lungs, distribution of saturated blood to body tissues through arteries and capillaries, burning of carbohydrates in tissue cells-producing energy, return of oxygen depleted blood and waste, like carbon dioxide gas, back to heart via capillaries and veins, exhaling of carbon dioxide gas, keeps going on continuously in the human body.

From the foregoing discussion, it is clear that:

1. There are arteries and associated capillaries carrying oxygen-saturated blood, veins and associated capillaries filled with oxygen depleted blood and tissues with carbohydrate filled cells in every part of the human body.
2. The **amount of blood in arteries and veins keeps** ***fluctuating with time***. The amount of blood in arteries increases each time it is pushed by the pumping action

of the heart and decreases thereafter with the pulse rate. The oxygen depleted blood and waste materials like carbon dioxide are carried by veins back to the heart. The amount of oxygen depleted, and waste loaded blood in veins also fluctuates with time but by a smaller amount and at a slower rate. It is because the pumping force generated by the contraction of the left ventricle of the heart diminishes by the time the oxygen depleted blood (loaded with waste) reaches veins. In the third component of the body, the tissues and cells, the amount of blood remains nearly constant with time.

3. Oxygen level in the oxygenated blood is very important because it regulates the production of energy in cells.

The level of saturation of oxygen in blood may alter because of several reasons including the quality of the air inhaled and the malfunctioning of lungs. It is, therefore, required to monitor the oxygen saturation level in a patient at regular intervals; oxygen saturation level of blood is now treated as one of the essential parameters, more so after the COVID pandemic which induced in most of its patients a sharp fall in level of oxygen saturation of blood.

In principle, there are two methods to find the oxygen saturation level of blood: (i) to draw some blood from an **artery** (mostly from the wrist artery) of the patient and then carry out detailed analysis of the blood sample in a laboratory. This method, called a***rterial blood gas test (ABG)*** gives more accurate results and provides many other useful information's about the blood but it is time consuming, it may take a few hours. (ii) The oxygen saturation level of the blood may be obtained almost immediately by using a pulse oximeter.

A pulse oximeter is a small device (see Fig. 3.2) that is put on some thinner part of the body, mostly on the finger of the patient but may also be put on the earlobe. As may be observed in this figure, the oximeter looks like a clip with two flaps between which the finger of the patient is held. The working of the oximeter is based on two facts: (a) If oxygen-saturated blood (flowing in arteries) is illuminated by a source of **infrared** (IR) radiations, it absorbs a large amount of these radiations and allows only a small amount of IR radiations to pass through. Further, the amount of IR radiation absorbed (reduction in the intensity of IR) is proportional to: (i) the **oxygen saturation level** of the blood and (ii) the **amount of the blood**. In contrast, the absorption of **red-light** in oxygen-saturated blood is very small and it allows a large amount of red-light to transmit through it. (b) On the other hand, the oxygen depleted blood (flowing through veins) absorbs a large amount of red-light, allowing only a small intensity of red-light through. The amount of red-light absorbed by oxygen depleted blood is proportional to the **amount** of the depleted blood. Further, the oxygen depleted blood absorbs only a very little amount of IR, allowing large amounts of IR radiation through. It may be remarked that the IR and the red-light absorption properties of the two types of the bloods are just opposite of each other.

A pulse oximeter has two types of light sources: one that emits red-light of wavelength 660 nm (nanometer, $1 \text{ nm} = 1 \times 10^{-9}$ m) and the other that emits infrared light (IR) of 940 nm. These light sources are very small like the LED lights used for decoration purposes. The two light sources are attached to one of the flaps of the clip

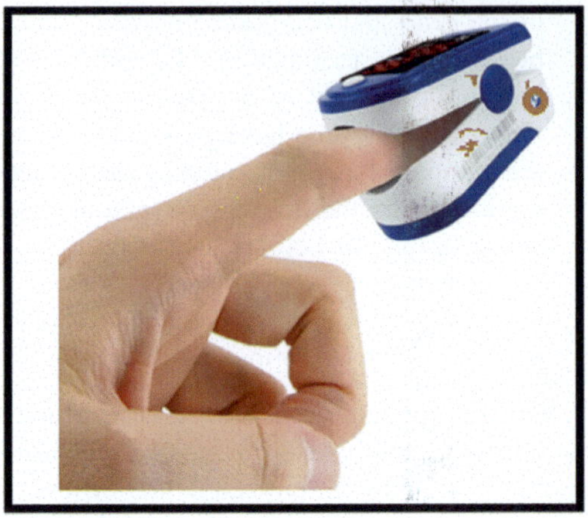

Fig. 3.2 Image of a pulse oximeter

like-oximeter. There is also a tiny detector that may measure the intensities of both the red-light and the IR radiation. The detector, which is a photodiode, is fixed in the other flap of the device, opposite to the two light sources. When the oximeter is worn on the finger, the light emitted by either of the two sources passes through the finger and the detector records the intensity of the transmitted light. There is also an A/D converter, a microcomputer with memory, an LCD screen and small ion-battery as a power source, included in the body of the device. Figure 3.3 shows the layout of a typical pulse oximeter.

The instrument manufacturer carries out the calibration of the pulse oximeter before putting it in the market for sale. In the calibration stage of the device, the two light sources of the oximeter are switched on, one after the other, each for a predetermined short interval of time '*t*' (of the order of 0.5 s), when there is nothing between the flaps of the instrument. Electrical signals produced at the output of the photodiode, corresponding to the two sources, are digitized using the A/D converter and are recorded as the initial intensities, I_0^{IR} and I_0^{Red}, respectively of the infrared (IR) source and the red-light source. In the next stage of calibration

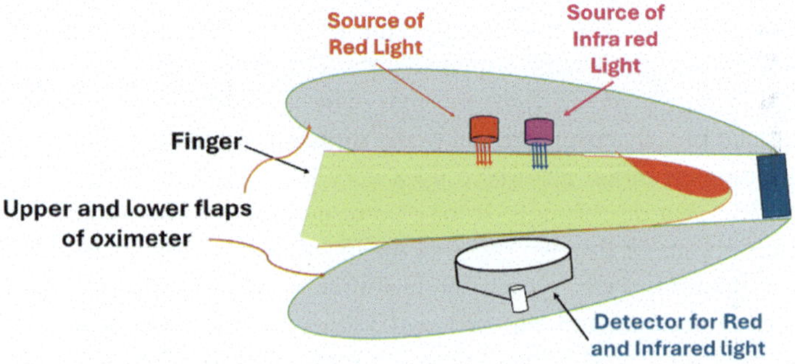

Fig. 3.3 Layout of a typical pulse oximeter

fingers of volunteer patients with known blood saturation levels, say *a*, *b*, *c*, …., etc., are held between the flaps of the oximeter one patient after the other, and the two light sources are switched on one after the other for the same time '*t*'. The corresponding pairs of digitized intensity readings, $I_a{}^{\text{IR}}$ and $I_a{}^{\text{Red}}$; $I_b{}^{\text{IR}}$ and $I_b{}^{\text{Red}}$; $I_c{}^{\text{IR}}$ and $I_c{}^{\text{Red}}$; …, etc., are recorded for each level of oxygen saturation. The computer, inbuilt in the device, then calculates the value of the ratio *R* (given below) for each value of saturation.

$$R_a = \left(I_a^{\text{IR}} \times I_0^{\text{Red}}\right) / \left(I_a^{\text{Red}} \times I_0^{\text{IR}}\right) \tag{3.1}$$

A graph between the saturation levels $a, b, c\ldots.$ versus the ratio values $R_a, R_b, R_c\ldots$, etc., provides the calibration curve which is stored in the memory of the computer for later use. To determine the oxygen level in the blood of a patient, a finger of the patient is first cleaned by a suitable cleaner and is then put between the two flaps of the oximeter. On switching-on the oximeter, the infrared and red-light sources in the device shine one after the other for the fixed period '*t*'. The digitized intensities $I_P{}^{\text{IR}}$ and $I_P{}^{\text{Red}}$ get generated from the electric signals produced by the photodiode sensor and processed by A/D converter. The processing unit of the inbuilt computer then calculates the ratio R_P of the patient as defined by Eq. 3.1. The value of oxygen saturation of a patient's blood is then obtained using the calibration curve already stored in the memory of the computer. All these steps—recording of infrared and red-light intensities, their digitization, calculation of the ratio R_P and extrapolation of patient's saturation level from the stored calibration graph, etc.—are done within a fraction of a second by the inbuilt computer. The oxygen level as obtained by these measurements is then displayed on the LCD screen.

One may ask what is the significance of the ratio *R*? It has been found that the ratio as defined by Eq. 3.1 is stable, does not depend on the pulse rate of the patient and is normalized with respect to the oxygen level of the blood in the veins of the patient. In some more accurate oximeters, the effect of the absorption of IR and Red-light in tissues of the finger is taken in account by recording the intensity of light by the photodiode detector when both the light sources are off, and the detector is on for time '*t*'. A slightly modified form of ratio *R* is then used for calculating the saturation level.

Oxygen saturation level of blood is expressed in percentage (%). For a healthy person, the oxygen saturation level of the blood lies in the range of 95–100%. Saturation level below 95% is called ***hypoxemia*** and may be a cause for concern. Saturation level above 100% is rare and may occur when a patient is kept on an oxygen mask for a considerable long time in a hospital.

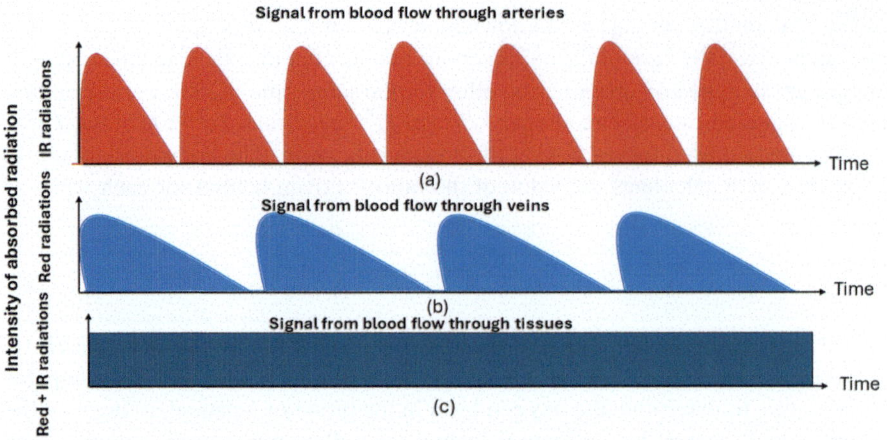

Fig. 3.4 Time variation of the intensity of **a** infrared radiations absorbed by oxygen-saturated blood passing through arteries **b** red-light radiations absorbed by the oxygen depleted blood passing through veins **c** total radiations (IR + Red) absorbed by the tissues of the body part

3.10 Measuring the Pulse Rate

An oximeter also measures the pulse rate of the patient; that is why it is called the pulse oximeter. Information about the pulse rate is derived from the time variation of the amount of blood through arteries. As has been mentioned, the absorption of IR radiation in the finger of the patient is proportional both to the amount (or volume) of the oxygen-saturated blood and its saturation level. The detector (photodiode) in the oximeter records the time variation of the intensity of the IR radiation absorbed by the finger ($I_P{}^{IR} - I_0{}^{IR}$). This time variation graph looks like the one shown in Fig. 3.4.

Figure 3.4a shows the time variation of the intensity of IR radiation absorbed by the oxygenated blood flowing through the arteries of the body part (finger) held between the flaps of the oximeter. As already mentioned, the intensity of the absorbed IR radiation depends on two factors: the level of oxygen saturation and the amount of blood. If the level of oxygen saturation of the blood passing through the arteries remains constant, the time variation of the IR absorption intensity may be assigned only to the variations in the amount of the blood. The obvious question that arises now is why does the amount of blood passing through arteries change with time? The answer is simple: at each beat of the heart, when the muscles of the left ventricle of the heart contract, blood is directly pushed into arteries. As such between two successive heart beats, the amount of blood in arteries increases to the maximum value and then decreases. Heart beats are produced by the rhythmic contraction and expansion of heart walls. Contraction of the walls of the heart chamber exerts high pressure on the oxygenated blood pushing it into arteries increasing the amount (or volume) of blood through them, while during the relaxation of chamber walls pressure on the blood gets reduced. It, therefore, means that the heartbeat rate (number of times heart beats in a minute) is equal to the number of variations in the IR intensity absorption graph (shown in Fig. 3.4) at the same time of 1 min. The time variation of absorption intensity is converted into corresponding electrical signals by the photodetector of the oximeter. The associated electronics of the device count the frequency of these electronic pulses which is equal to the pulse rate of the patient. For a normal healthy person, the pulse rate may vary between 60 and 100 beats per minute, depending on age and other factors.

Figure 3.4b shows the time variation in the intensity of the red-light absorbed in deoxygenated blood flowing through the veins of the patient under observation of the oximeter. It may be remarked that the absorption intensity variations in this case are not as pronounced as in the case of arteries. Further, the frequency of these variations is also smaller than the case of the arteries. It is because the deoxygenated blood in veins does not get a direct push from the heart when the heart contracts. The effect of heart contraction reaches veins indirectly via arteries, body tissues and interlaying cells; that considerably dilutes the push, reducing the absorption intensity and repetition frequency.

In Fig. 3.4c, which is for the tissues of the body part of the patient under the observation of the oximeter, the absorption intensity is very low and there is no time dependent component of absorption intensity. Body tissues, on average, have nearly the same amounts of the oxygenated and the deoxygenated blood that remains nearly constant with time.

The electronic circuitry in the oximeter selects the fast electronic signal generated by the flow of oxygenated blood through arteries, shown in Fig. 3.4a, and rejects the low frequency signal generated by the flow of deoxygenated blood through veins (Fig. 3.4b) and the constant intensity signal (Fig. 3.4c) due to tissues and cells. The frequency of the signal from the arteries gives the pulse rate or the heartbeat rate.

Chapter 4
The Sphygmomanometer

Abstract Heart pumps oxygenated blood in arteries of the human body at regular intervals. Blood flowing through the arteries exert pressure on the walls of these arteries; this pressure is called blood pressure or Bp. The blood pressure may have two values, higher (or systolic) when the heart has just pumped and the lower (diastolic) just before the next cycle of pumping. Both these Bp values are important from the health point of view. The science behind the measuring of the two Bp values is discussed in this chapter.

Blood Pressure: We are conversant with the term 'atmospheric pressure'; the hydrostatic pressure exerted by the weight of atmospheric gases (gases contained in the atmosphere) on the earth. Atmospheric pressure is defined as the force exerted by the atmospheric gases per unit area of the earth surface. Though pressure may be measured in several different units, like the mm of Hg (Mercury), barn, Pascal, etc.; however, we shall limit at present to the unit 'millimeter of Mercury', written in short as ***mmHg***. For example, the value of atmospheric pressure at sea level under normal conditions of temperature, etc., is 760 mmHg, which means that a mercury column of height 760 mm will exert the same pressure as is exerted by the atmospheric gases at a place on earth surface at sea level.

Human heart pumps oxygenated blood to different body parts through arteries. The pumping action of the heart takes place by the contraction of the walls of the left ventricle of heart that exerts pressure on the oxygenated blood contained in it. The oxygenated blood running through arteries, in turn, exerts pressure on the artery walls. Pressure exerted by the oxygenated blood on the walls of arteries is called 'blood pressure' in short written as bp, Bp (or BP). For a given person there may be two (maximum and minimum) values of the bp; pressure felt by the artery wall when the left ventricle of heart contracts—this ***maximum*** value of blood pressure is termed as the ***systolic*** bp value. The pressure of blood felt by the artery wall when the left ventricle of heart reflexes is termed as ***diastolic*** bp; this is the ***minimum*** value of the blood pressure for a particular person. It is obvious that for any person the systolic bp is higher than his diastolic bp. Knowledge of both the systolic and the diastolic bp values of a patient is required as they play a crucial role in evaluating

R. Prasad, *Understanding Medical Devices*, Series in BioEngineering,
https://doi.org/10.1007/978-3-031-88630-0_4

the health of the patient's heart. It is a common practice to write the two bp values as: (Systolic bp value/diastolic bp) mmHg. For example, a healthy person may have bp values as 120/80 mmHg, meaning thereby that his systolic bp is 120 mmHg and diastolic bp as 80 mmHg. The upper or systolic bp for a healthy person may range from 90 to 140 mmHg and the lower or diastolic from 80 to 60 mmHg. Values of bp above 140 mmHg for systolic and above 90 mmHg for diastolic generally refers to high blood pressure, while systolic bp below 90 mmHg and diastolic below 60 mmHg are categorized as low blood pressure.

The scientific name of the instrument used to measure blood pressure is sphygmomanometer; pronounced as: *SFIG-moh-ma-NO-mi-ter*. The instrument was invented in 1861 by Samuel Siegfried Karl Ritter von Basch, an Austrian physician. The instrument has undergone several improvements since its invention incorporated by several scientists/physicians including Scipione Riva-Rocci (Italian pathologist/pediatrician), Harvey Cushing (USA neurosurgeon) and Nikolai Korotkov (Russian physician).

Sphygmomanometers may be of three types; (a) basic device that uses Mercury column (b) the aneroid bp instrument and (c) the digital bp unit.

a. **Basic bp Measuring Device with Mercury Manometer**

All bp measuring instruments consist of four basic components; (i) an ***inflatable arm cuff*** that contains a rubber bladder in which air can be pumped, (ii) a **rubber *bulb like pump*** for pumping air into the bladder of the cuff to increase the air pressure in the bladder, (iii) a knob/screw like ***valve** at the top of the rubber bulb*, opening of which allows the release of air from the cuff bladder in a controlled way so as to control the air pressure in the cuff and (iv) a ***device to record the pressure*** of air in the cuff at different settings. In basic (manually operated) bp instruments, a column of mercury (mercury filled in a capillary tube) with marks corresponding to different pressure values (in mmHg units) is used. Apart from the above-mentioned components, the basic instrument also requires a stethoscope. Figure 4.1a shows the different components of the manual mercury manometer bp instrument along with the stethoscope.

A rough sketch of the method of recording bp of a patient using a manual mercury instrument is shown in Fig. 4.1b. The cuff, in most cases, is tied some 5 cm above the elbow on the upper forearm of the patient. However, in some special circumstances, the cuff may also be tied around the wrist or some other suitable body part. It is important to use the cuff of the correct size suitable for the body part where it is wrapped. The advantage of using the upper forearm is that the ***branchial artery,*** the main artery that carries blood to different arteries of the arm passes through the middle of the upper forearm and it is easy to detect the changes in the flow pattern of blood through this artery using a stethoscope. Blood flowing through an artery without any constriction or hindrance makes no sound; however, if the blood flow is restricted by any means (like the narrowing of artery by the application of excessive pressure, etc.), the blood flow through the artery makes a typical sound called ***Korotkoff sound.*** Korotkoff sound is produced by the vibration of artery walls due to the force applied by the blood in gushing through the constriction. Korotkoff

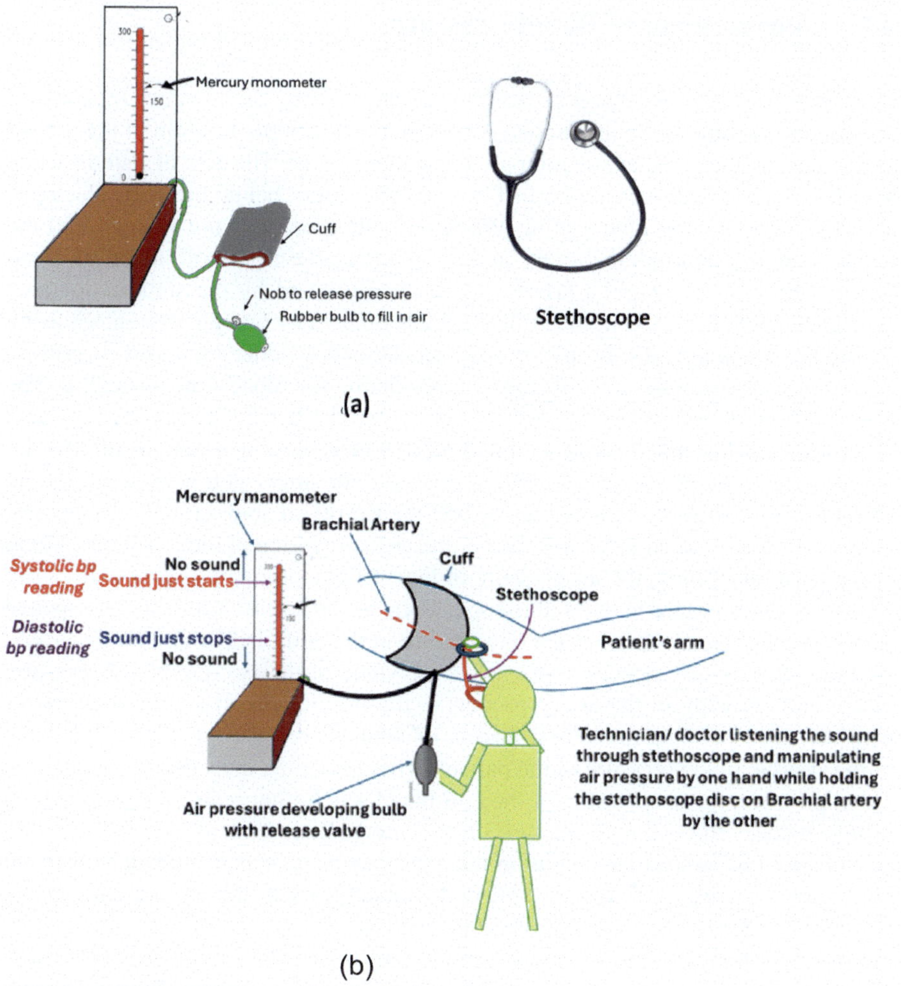

Fig. 4.1 **a** Diagram of the basic bp instrument with mercury manometer, **b** a rough sketch of taking bp measurements using manual mercury bp instrument

sound can be easily detected by a stethoscope when placed over or near the artery. Korotkoff sounds may vary in quality from tapping, swooshing, muffled sound to silence depending on the external pressure exerted on the artery to restrict the free flow of blood.

4.1 Measurement of Blood Pressure

In order to measure the blood pressure of a patient using the mercury manometer instrument, the cuff of the instrument is wrapped on the upper forearm of the patient of about 5 cm above the elbow joint as shown in Fig. 4.1b. The disk of a stethoscope may now be pressed over the part of the brachial artery below the cuff, placing it about 2 cm above the elbow joint. When the cuff bladder is not inflated and the blood is passing through the artery in a normal way, the Korotkoff sound of blood flowing through the brachial artery can be clearly heard by the stethoscope as a regular tapping sound. Now, the cuff bladder is inflated by pumping air into it using the rubber air pump. As the air pressure in the cuff bladder increases, it presses the brachial artery under the cuff bladder, restricting the blood flow through it. The Korotkoff sound received by the stethoscope changes both in intensity and frequency; it becomes muffled and swooshing. Ultimately, when the air pressure in cuff bladder exceeds the value of the systolic blood pressure, the artery gets totally closed, no blood flows through the brachial artery and the Korotkoff sound totally disappears, there is silence in the stethoscope. Therefore, as the first step of measuring the blood pressure of the patient, the air pressure in the cuff bladder is increased to the level when no sound is heard in the stethoscope. Then, the air release valve knob is opened slowly which reduces the external air pressure in the cuff bladder and on the artery. As the cuff pressure reduces and reaches the value of the systolic blood pressure, blood starts flowing in the artery producing the first Korotkoff sound; manometer pressure corresponding to the ***first Korotkoff*** sound is recorded and gives the value of the systolic blood pressure of the patient. After recording the value of the systolic blood pressure, the air pressure of the cuff bladder is further reduced slowly, which in turn reduces the external pressure on the artery. Reduction of external pressure on the artery allows blood to flow in the artery with lesser constraint reducing the pitch and intensity of Korotkoff sound. Ultimately, when the air pressure in the cuff bladder equals the diastolic blood pressure of the patient, blood starts to flow through the artery without any restriction and does not produce any Korotkoff sound. The manometer pressure corresponding to the instant when ***Korotkoff sound totally disappears*** while releasing the air pressure of the cuff bladder gives the value of the diastolic blood pressure of the patient. It is obvious that recording of the systolic and the diastolic blood pressures using a mercury manometer instrument requires training of the technician/doctor to identify the points where the Korotkoff sound first appears and the point where this sound totally disappears in the deflation phase. The art of understanding the Korotkoff sounds is called ***auscultation***, and bp measurement by mercury column sphygmomanometer requires a doctor/technician who is good at auscultation. Mercury column-based sphygmomanometers are referred to as the ***golden bp instruments*** as they are very accurate and are used for the calibration of other types of bp instruments. The one and rather biggest disadvantage of these instruments is the use of mercury which is a health hazard on account of its toxicity. It is known that all substances, including solids, have their vapors around them. Mercury vapors, if inhaled, can cause permanent lung damage and potential brain

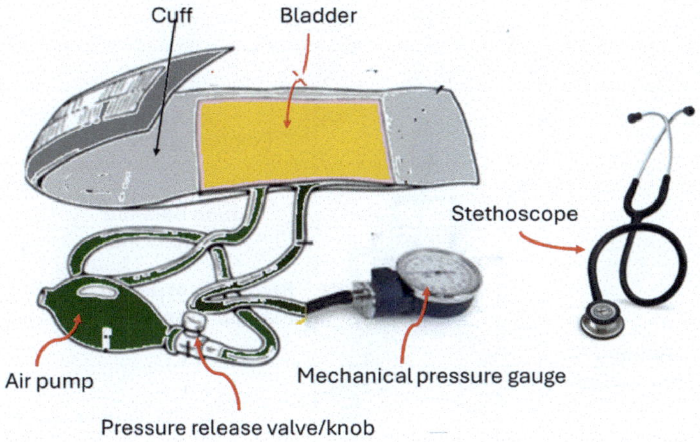

Fig. 4.2 Components of an aneroid sphygmomanometer

damage. It may also damage kidneys and the central nervous system; long exposure may lead to death if not treated.

b. **Aneroid Sphygmomanometer**

An aneroid bp instrument, that does not contain any fluid like mercury, has an inflatable cuff with a bladder, a rubber air pressure pump with air release valve, a mechanical pressure gauge (in place of the mercury manometer) attached to the cuff and a stethoscope, as shown in Fig. 4.2

As may be observed, the mercury manometer of the mercury bp instrument has been replaced by a mechanical pressure gauge in the aneroid (without any fluid) bp instrument. The operation of the aneroid instrument is similar to the mercury manometer bp instrument; a stethoscope is still used to monitor the Korotkoff sounds to register the systolic and diastolic bp readings. The corresponding pressure values are now read on the circular dial of the mechanical pressure gauge (instead of a mercury manometer).

An analog mechanical or aneroid pressure gauge consists of a flexible metallic (or a ceramic) surface that deforms under pressure; deformation is being proportional to the pressure. The pressure-induced deformation is amplified using mechanical gear and lever systems that make a pointer to rotate on a circular scale. The circular scale of the gauge is calibrated in mmHg units using a conventional mercury manometer. Aneroid pressure gauges with several designs of metallic pressure sensors, like the diaphragm metallic surface, metallic bellows, single bellow, double bellows, etc., are available. The working of a dual bellows aneroid pressure gauge that measures the absolute pressure may be explained with the help of Fig. 4.3

As shown in this figure, Bellow-I, housed in a cylindrical container, is evacuated and sealed so that its top surface remains fixed and does not flex. The container volume of Bellow-II is connected to the bladder of the cuff so that it is filled with the air at the pressure of the cuff bladder. Since the bladder pressure is higher (than

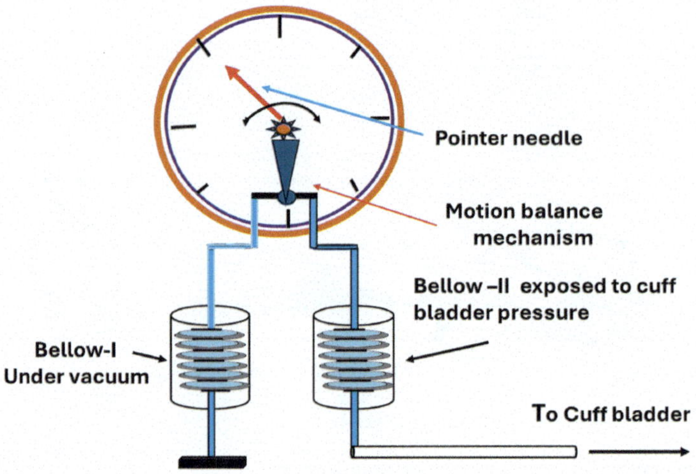

Fig. 4.3 Sketch of an aneroid twine bellows pressure gauge

the pressure at the top of Bellow-I, vacuum pressure that is zero), the top surface of Bellow-II bulges out, bulge being proportional to the pressure in the bellow. The outer surfaces of the two bellows are connected by a motion balance mechanism that rotates the pointer needle on the circular scale. The circular scale may be calibrated, in an auxiliary experiment, in terms of mmHg by filling the Bellow-II with air at known pressures. Since the pressure measurement in this case is done against the bellow that has vacuum, the measurement gives the absolute pressure.

A diaphragm-type aneroid pressure gauge employs a thin flexible metallic diaphragm that divides a metallic chamber in two parts. The lower part of the chamber is connected to the volume (like the bladder of the cuff) the pressure of which is required to be measured. The upper part of the chamber remains connected to the atmosphere through a small hole. Since the pressure on the cuff bladder is more than the atmospheric pressure, the diaphragm flexes up words producing a bulge or bubble on the surface. The bulge of the surface, which is proportional to the pressure difference between the atmospheric pressure and the bladder pressure, is amplified using a lever system and is made to rotate a needle on a circular scale. The circular scale is calibrated in terms of mmHg using a mercury manometer. Sketch of a diaphragm-type aneroid pressure gauge is shown in Fig. 4.4

c. **Fully Automatic Digital Display Sphygmomanometer:**

Manual auscultatory measurement of upper arm blood pressure with mercury sphygmomanometer was the gold standard non-invasive test and mainstay clinical method to diagnose hypertension in the twentieth century. Fully automatic bp measuring instruments become favored over the manual methods, including the aneroid instruments, to lessen the chances of user errors from such things as incorrect stethoscope placement, incorrect cuff size, observer bias, failure to correctly interpret Korotkoff sounds, etc. Also, it avoided the use of mercury that is considered a toxic health

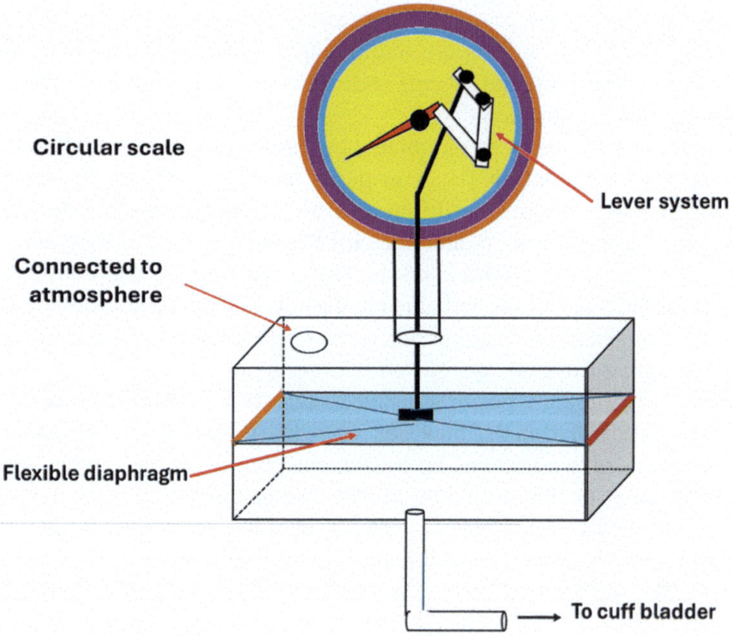

Fig. 4.4 Sketch of a diaphragm-type aneroid pressure gauge

hazard. Digital display of the systolic (upper) and the diastolic (lower) blood pressure readings is an added advantage of these automatic bp instruments. The first commercial automatic bp measuring device became available in 1976. A major reason for the rise in the use of automated bp instruments is the global policy directive in 2005 to phase out and replace mercury-based bp instruments in health care settings due to toxicity concerns (Fig. 4.5).

In auscultation instruments use is made of the sound (Korotkoff) that is produced by the vibrations of the artery walls when blood rushes in the artery through some obstruction. In automatic bp instruments, the oscillations produced in the artery walls due to the rushing of blood in constrained artery are picked up. Oscillating artery walls ***induce similar oscillations in the air contained in the cuff bladder***.

Fig. 4.5 Automatic BP instrument

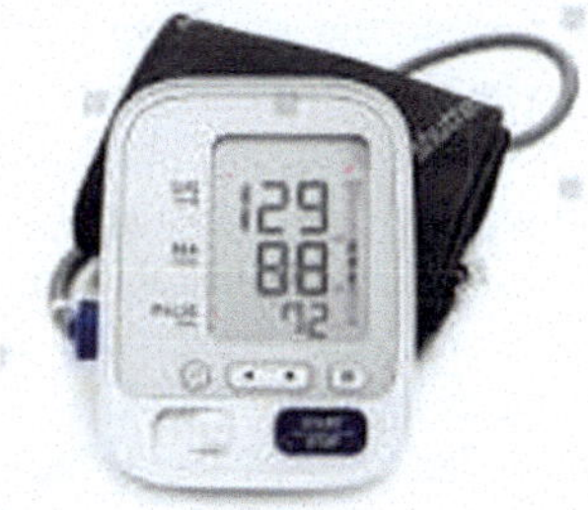

These air oscillations are converted into electric/electronic signals using suitable transducers. These instruments are traditionally called '**oscillometric**' bp devices since the waveforms recorded by cuff and subsequently analyzed for the estimation of bp are oscillatory pulsations.

Essential components of an automated bp instrument are: (a) a **Microcomputer processing unit** that controls (i) an air pump that may pump air into cuff bladder, (ii) a **control valve** that may build up pressure in the cuff up to a desired level in the inflation phase and may reduce the cuff bladder pressure slowly in small steps of, say, 3 mmHg in the ***deflation phase***, (iii) a **transducer** that converts pressure oscillations induced in the air contained in the cuff bladder (due to the oscillations of artery walls) into electronic signals and (iv) a **LCD** (Liquid Crystal Display) **Screen** to show the digital values of systolic and diastolic pressures.

In principle, different types of transducers may be used to convert pressure oscillations into electrical signals, however, transducers based on ***Piezoelectric effect*** are commonly used in automatic bp instruments. Some natural crystals, like quartz, when cut along a specific direction, acquire the property that if pressed along one direction, and they develop electric charge, in some other specific direction. The amount of charge is being proportional to the exerted pressure. Since small piezoelectric crystals are easily available at low cost, this technique is often used for the conversion of pressure variations into electrical signals. Another technology, though rarely used, is the ***Capacitive Diaphragm Technology***. Capacitive diaphragm pressure transducer uses a flexible diaphragm made of a conducting material between two electrodes. When pressure is applied to the diaphragm, it deflects causing a change in its electrical capacity which may be converted into an electric signal.

The automatic bp instrument essentially operates in the deflation phase, when the air pressure in the cuff bladder has been raised slightly above systolic value and then the pressure is being slowly reduced in small steps. The recorded cuff bladder deflation curve has two components; (a) a slowly declining component as the cuff bladder pressure decreases and (b) the pulsations caused by the vibrations of artery wall (the cardiac contraction and relaxation). The microcomputer carries out the analysis of the electrical signals produced in the deflation phase and generates the envelope of the electronic pulse size. This envelope is called the ***Oscillometric waveform***.

Figure 4.6 shows the pressure oscillation waves generated in the cuff bladder during the deflation phase (Fig. 4.6a) and the corresponding electronic signal waves produced at the output of the transducer (Fig. 4.6b). The microcomputer of the device carries out the analysis of the oscillometric envelope (Fig. 4.6c) and determines the ***maximal value pressure*** point on the curve. It has been found that the maximal value pressure is related to ***both the systolic bp value and the diastolic bp value*** through some empirical relations. The computer algorithm generates the analog values of the systolic and diastolic bp values corresponding to the obtained maximal pressure value using the prescribed empirical relations. An A/D converter within the device converts analog values of systolic and diastolic bp values into digital form which are displayed on the LCD screen.

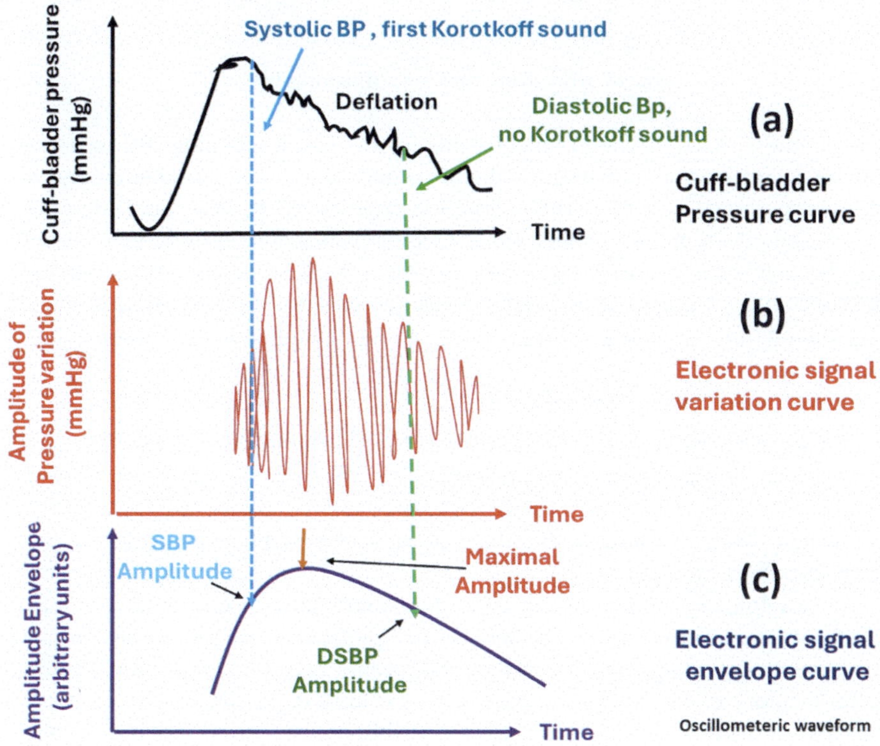

Fig. 4.6 Oscillometric waveform analysis in automatic BP instruments

In a fully automatic bp instrument, the microcomputer within the device controls the buildup of pressure in the cuff bladder to a value slightly above the systolic bp of the patient and then slowly deflates the bladder. In the deflation phase, the microcomputer also records the spectrum of the electronic signals produced in the output of the transducer that converts pressure variations into electronic signals. The electronic pulse structure delivered by the transducer contains a mixture of two types of signals, (i) corresponding to the vibrations of the artery wall and (ii) the signal corresponding to the pressure variation due to the deflation phase. The microcomputer separates the electronic signals corresponding to the vibrations of the artery wall from the electronic signals associated with the decreasing pressure due to the deflation phase; further, it retains signals corresponding to the artery wall vibrations and rejects the other. Next, the microcomputer generates the envelope of the electronic signals corresponding to the vibrations of the artery wall and determines the **maximal amplitude** $\boldsymbol{A_m}$ of the envelope curve. It is important to note that, in general, the envelope curve has different shapes for different patients and consequently different magnitudes for maximal value. It has been observed that the maximum amplitude A_m is related to the magnitudes of the systolic bp and the diastolic bp values through some algebraic relation. For example, it may be assumed that the relationship is of the following type'

$$A_m = B(\text{systolic bp}) + C(\text{diastolic bp}). \tag{4.1}$$

Here B and C in Eq. 4.1 are two constants for the given automatic BP measuring unit. To determine the magnitudes of the two constants, (A and B), an auxiliary calibration experiment is done. In this calibration experiment, the maximum amplitudes $A_{m1}, A_{m2}, A_{m3}, \ldots$, are determined for several patients whose systolic bp's (systolic bp)$_1$, (systolic bp)$_2$, (systolic bp)$_3$ and diastolic bp's (diastolic bp)$_1$, (diastolic bp)$_2$, (diastolic bp)$_3$ are already known (from measurements by a mercury BP instrument). Substitution of these values in Eq. 4.1 gives a set of simultaneous equations,

$$\begin{aligned} A_{m1} &= B(\text{systolic})_1 + C(\text{diastolic})_1, \\ A_{m2} &= B(\text{systolic})_2 + C(\text{diastolic})_2 \quad (\text{Set of simultaneous equations}), \\ A_{m3} &= B(\text{systolic})_3 + C(\text{diastolic})_3. \end{aligned}$$

The best values of constants B and C may be determined by solving the above set of simultaneous equations.

The example of the relationship that we have taken above is just for illustration; in actual case, the relationship may be far more complicated involving several constants. The computer language version of the relationship, called the algorithm, is stored in the memory of the microcomputer which readily determines all the unknown constants of the relation from the data of the calibration experiment and stores them in its memory. With the calibration done, the instrument is ready to measure the blood pressure values for an unknown patient. An analog-to-digital converter inbuilt in the system delivers the digital values of the two blood pressures which are displayed on the LCD screen.

A flowchart of processes in an automatic bp instrument, which is self-explanatory, is shown in Fig. 4.7.

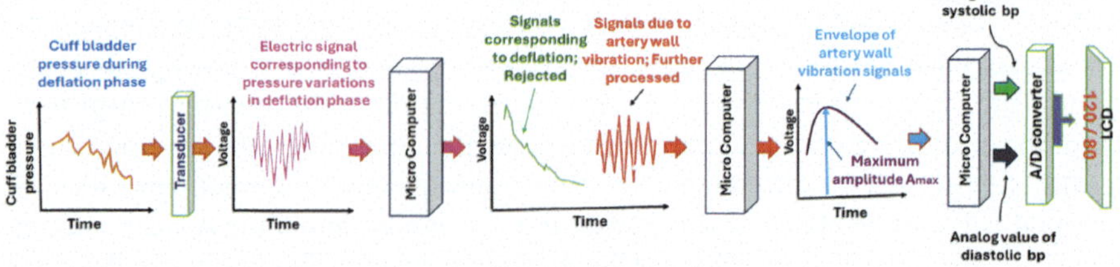

Fig. 4.7 Flowchart of processes in an automatic sphygmomanometer

Chapter 5
X-Ray Imaging

Abstract X-rays are electromagnetic radiation of a specific frequency range and are frequently used to take images of damaged/broken bones, ligaments, etc. The science behind production and image formation using X-rays is discussed in detail in this chapter. The role of image intensifier along with the quality of X-ray image is also discussed.

5.1 Introduction

X-ray and ultrasound imaging are the two powerful tools to assess the health and the physical condition of internal body organs. Both the X-rays and the ultrasound rays are waves, but the nature of these two kinds of waves is quite different. X-rays are electromagnetic waves, like the waves or rays of visible light; however, X-rays are not visible to our eyes. It is because human eyes are not capable of detecting X-rays, in other words, the human eye is not a good detector for X-rays. In a way it is a limitation of our eyes; there are other detectors that may detect and measure the parameters like the wavelength, intensity, etc., of X-rays. Like visible light, X-rays may travel through vacuum; it does not require any medium to travel. Sunlight, which contains both visible and invisible components of electromagnetic radiation, reaches Earth after crossing a large length of vacuum. On the other hand, ultrasound waves are mechanical waves, and they may be created only in an elastic medium like air, liquid or solids. Ultrasound waves also require an elastic medium to travel. Ultrasound waves are sound waves of higher frequencies. Like all waves, both the X-rays and the ultrasound waves get reflected, refracted and absorbed when they hit the boundary between two different media. In this section we shall discuss the properties and applications of X-rays in some detail.

X-rays were discovered accidentally, by Wilhelm Conard Roentgen, Professor of Physics in Wurzburg, Bavaria, Germany, (a German by birth who grew up in Holland), in the year 1895 after seven weeks of assiduous work. During this period Roentgen studied the properties of this mysterious radiation and found that X-rays may penetrate through screens of notable thickness. Since the exact nature and other

R. Prasad, *Understanding Medical Devices*, Series in BioEngineering,
https://doi.org/10.1007/978-3-031-88630-0_5

properties of these radiations were not known at that time, Roentgen named them X-rays. Discovery of X-rays immediately revolutionized the medical diagnostic world, as early as January 1896 Physicians started using X-rays to photograph the skeleton of their patients for detecting bone fractures and subsequently their lungs for TB and other infections.

5.2 X-Ray Machine

As already mentioned, X-rays form a small component of frequency range 3×10^{16}–3×10^{19} Hz (that corresponds to the energy range of (100 eV–100 keV) of the vast spread of electromagnetic radiation spanning from Gamma rays to the radio waves. In general, electromagnetic radiation of different frequencies is generated when a lightweight electrically charged particle, like an electron, either decelerates or accelerates. Since the phenomena of the emission of electromagnetic radiations of wide frequency spread on declaration of lightweight charged particles were first observed by German scientists, these radiations were called '***Bremsstrahlung***' a German language word which in English means the Braking radiations, radiations produced by stopping or deceleration of charged particles.

X-rays, like all electromagnetic waves, exhibit both the ***particle and the wave nature***; in some experiments they behave like waves and in others as particles. The particle associated with electromagnetic waves is called the ***photon.*** It is important to note that when an electromagnetic wave behaves like a wave, it does not exhibit its particle nature and vice versa. The dual nature of electromagnetic waves may be looked at as the two sides of a coin; when one looks at the head of the coin, the tail of the coin is not visible, and while looking to the tail, the head of the coin is not seen. Because of this dual nature, the electromagnetic radiations, including the X-rays, may be specified in terms of its wavelength (denoted by λ) or the energy of its photon E (=$h\nu$; here h is Planck's constant and $\nu = c/\lambda$ is the frequency of the photon; c is being the speed of light in vacuum) in units of electron volt (eV). X-rays may be broadly classified into two types; (i) the ***hard X-rays*** that have photons of energy above 5 keV [wavelength below 0.1 nm (nanometer = 1×10^{-9} m)] and **soft X-rays** with photons of lower energy ($E < 5$ keV and wavelength higher than 0.1 nm). Hard X-rays may penetrate through a relatively thick layer of matter, while the soft X-rays get easily stopped by a thin layer of the matter.

5.3 X-Ray Machine

X-rays are emitted when high-speed electrons are made to hit (and slow down or come to stop) a target of some heavy metal like tungsten. There are three basic components of an X-ray machine; (a) an electron gun, which provides a collimated beam of electrons, (b) a high-voltage power supply that generates DC high-voltage pulses with desired value of voltage and pulse duration, (c) a suitable target, called anode, generally made of high melting point material like tungsten. In modern machines, anode is kept at some angle with respect to the incident electron beam and is mounted on a rotor to avoid damage to it due to excessive heating. The electron gun and anode assembly are kept inside an evacuated chamber, covered by thick metal- glass walls that are strong and transparent. Basic layout of an X-ray machine is shown in Fig. 5.1a.

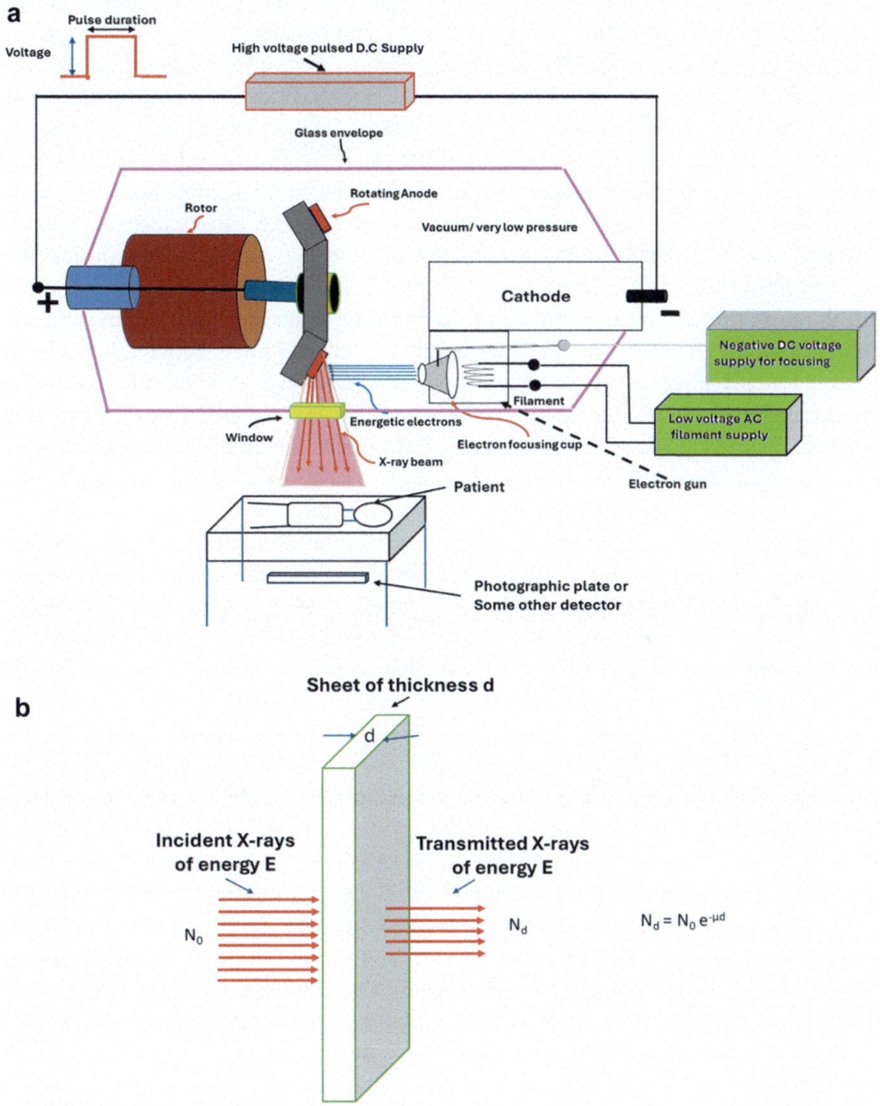

Fig. 5.1 **a** Sketch of an X-ray machine; **b** attenuation of monoenergetic X-rays while passing through a sheet of matter

The electron gun consists of a filament that is connected to a low-voltage AC power supply that provides the heating current. Heating current can be controlled to the desired value. Electrons in large numbers are emitted from the filament on heating by the process called ***thermionic emission***. The electron gun assembly is connected to the (cathode) negative terminal of the high-voltage pulsed power supply, the positive terminal of which is connected to the anode of the machine. Negatively charged electrons emitted by the filament are attracted by the positive potential at the anode and get accelerated. In order to focus electrons in a collimated beam, a hollow metallic cone is placed in front of the filament and a small DC negative voltage is applied to the cone. The negative potential on the cone pushes the electrons (emitted from filament) towards the axis of the cone producing a collimated beam of electrons which is accelerated by the positive voltage at the anode. An anode is a tungsten disk that is mounted on a rotor. Energetic electrons on hitting the anode come to rest and lose their kinetic energy (energy due to their motion) in the form of heat and X-rays. More than 98% of the energy of accelerated electron beam is converted into heat when it hits the anode target. Only a very small fraction of the electron kinetic energy is converted into X-rays. The anode disk makes a small angle with the direction of the incident electron beam to provide a focused and intense beam of X-rays. The cathode–anode assembly of the machine is kept enclosed in an evacuated metal–glass envelope which is hard and transparent. Vacuum or very low pressure inside the glass envelope is required so that fast electrons moving towards the anode do not collide with the molecules of the gas inside the envelope. If some air is present in the envelope, the electron beam will get scattered by the collisions with air molecules and electrons will be lost from the beam. There is a window, mostly of beryllium metal, fixed in the glass envelope to allow the X-ray beam to emerge out of the vacuum enclosure without much loss of its intensity. Absorption of X-rays in beryllium is minimal. A high DC voltage power supply that may provide voltage pulses of desired value of positive voltage and of time duration is used to provide positive potential to the anode of the device. The magnitude of the DC potential applied to the anode decides the maximum energy of the emitted X-rays. If V (volt) is the potential of the pulse applied to the anode, each electron of the incident beam will gain a kinetic energy of V electron-volts (eV, a unit of energy). When these electrons are stopped by the anode and their kinetic energy decreases, a part of the energy lost by electrons appears in the form of X-rays. As such an intense beam of X-rays is produced; the intensity of the X-ray beam depends on the intensity of the electron beam. The duration of the positive potential pulse provided by the power supply decides the duration for which the X-rays will be produced. As such, the maximum energy of the X-rays and the duration for which the burst of X-rays is produced may be controlled by setting the height and the duration of the positive voltage pulse. A considerable amount of heat energy along with X-rays is generated when high energy electrons strike the anode. If the electron beam is allowed to hit at the same point on the anode target, it may get burnt and damaged. To avoid this damage, the anode is made to rotate at a high speed, of the order of a few thousand revolutions per minute, with the help of a motor, rapidly changing the point of impact of electrons on the target. Target is additionally cooled in most machines by circulating cold water or transformer oil around it. In most X-ray machines, the height of positive voltage pulse may be varied from 15 to 140 kV (kilo volt). There may be three basic characteristics of an X-ray beam; (i) X-ray energy, (ii) intensity of the beam and (iii) the time duration of the for which X-rays are generated. In a commercial X-ray machine, all these parameters can be controlled to the desired value. The maximum energy of X-rays may be controlled by the voltage applied to the anode (i.e., the height of the anode pulse). The intensity of an X-ray beam is governed by the temperature of the filament; higher the filament current, higher will be the filament temperature which in turn will enhance electron emission and the strength (intensity) of the electron beam. The intensity or strength of an X-ray beam is directly proportional to the intensity of the electron current. The time duration of the X-ray burst may be controlled by the time duration of the voltage pulse. Though the kinetic energy (energy due to the motion) of all accelerated electrons is almost the same, the energy of emitted X-rays is not the same. It is because different amounts of energy are lost in the form of X-rays by different electrons of the beam. An electron that loses all its kinetic energy in the form of X-ray creates an X-ray of highest energy of V (eV).

X-rays may be broadly classified into two types; (i) the ***hard X-rays*** that have photons of energy above 5 keV (wavelength below 0.1 nm) and **soft X-rays** with photons of lower energy ($E < 5$ keV and wavelength higher than 0.1 nm). Hard X-rays may penetrate through a relatively thick layer of matter, while the soft X-rays get easily stopped by a thin layer of matter foil.

One of the characteristics of an X-ray beam that makes it useful for medical imaging is the penetrating ability of the beam. Let us consider a parallel beam of X-rays having say N_0 number of photons per unit area per second, each of the same energy E. Such a beam is called a ***monoenergetic*** beam with ***intensity*** N_0. When this beam is made to hit a screen or a sheet of some material of thickness, say d, then it is observed that the number N_d of X-ray photons of energy E that reach the other side of the sheet is smaller than the original number N_0. It means that $(N_0 - N_d)$ number of X-ray photons are lost by the incident beam while traversing the thickness of the material as shown in Fig. 5.1b.

Experiments have established the following relation between the intensity of the incident and the transmitted photon beams.

$$N_d = N_0 e^{-\mu d}. \tag{5.1}$$

Equation 5.1 says that the intensity of the transmitted photon beam decreases exponentially with the thickness d of the material. For example, for a given anatomic tissue, increasing the thickness of the tissue increases the attenuation of the X-ray beam, generally reducing the beam intensity to 50% for each 4 or 5 cm thickness. The quantity represented by symbol μ is called the ***linear absorption coefficient*** that represents the effectiveness of the medium to reduce the intensity of the incident X-ray beam. The magnitude of the linear absorption coefficient depends both on the nature of the medium (density and atomic number) and the energy of the X-ray photons. Relation 5.1 is strictly true for the case of monoenergetic X-rays; however,

X-rays produced by machines for medical purposes are generally not monoenergetic, and there are X-ray photons of different energies up to a maximum energy of V (eV); V is being the maximum voltage applied to the anode. Despite the energy distribution of X-ray photons, it is possible to obtain an average value for the linear absorption coefficient for a given material and for a given bunch of X-rays. Different components of anatomic objects, like muscles, tissues, ligaments, bones, etc., have different magnitudes for the average linear absorption coefficients for the given bunch of X-rays produced by a machine at a given settings of anode voltage and filament current. Therefore, different components of the anatomic object absorb/scatter and transmit different amounts of incident X-rays. The X-ray image is made by the transmitted fraction of incident X-rays.

The capability of creating images of patient's (anatomic) body parts by X-rays is because different body parts have different values of average linear absorption coefficients for the given bunch of X-rays. Therefore, different intensities of X-rays are transmitted by different body parts and organs producing different shades of gray in the X-ray photo image. For example, bones in the body have high density and, therefore, have the large value of average linear absorption coefficient. Bones absorb much more X-ray intensity as compared to soft tissues of low density, and hence, bones cast a very strong shadow in the X-ray image.

The intensity of X-rays ($N_0 - N_d$) lost in the thickness of the intervening body organ/part may be due to several effects; the X-ray might be absorbed by the body part, it may be scattered with nearly same energy in to some other direction (***called coherent or Rayleigh scattering***) or the X-ray may be scattered with reduced energy in some different direction (**called Incoherent or Compton scattering**). X-rays absorbed in the body are responsible for the health hazard, and they lose their energy in ionization of atoms and molecules in the body that may result in killing the body cells. Ionization is the process in which X-rays tear out electrons from the atoms of the body part. These electrons kill or damage the living cells of the part. Scattered X-rays form a part of the background noise and may deteriorate the quality of the X-ray image. Depending on the type of image receptor placed behind the patient's body part, there may be three types of X-ray images: (a) conventional film screen, (b) a photosensitive phosphor plate used in computed radiography or (c) a charged electron device used in digital radiography.

5.4 Conventional Film Screen

A normal photographic film is made of a cellulose acetate or polyester base with a coating of some light (visible) sensitive mineral like some silver salt. Initially cassettes of photographic films were used to record the X-ray images of anatomic objects in medical diagnostics. However, the images formed on such photographic films by X-ray exposure were very weak, not visible by the naked eye in a bright background. It was mainly due to two facts; (i) the films were not very sensitive to X-rays (films are more sensitive to visible light) and (ii) the intensity and flux of

X-rays used for taking the image are generally not sufficiently high. As a matter of fact, it is not desirable to expose a living body part to a high flux of X-rays on account of health hazards of excessive radiation dose. Soon, an additional piece of equipment called **'image intensifier'** was developed that enhanced the quality of X-ray images to a large extent.

Image intensifier Image intensifiers are electronic gadgets used to enhance the brightness or intensity of images. They are used to improve the quality of X-ray images of anatomic parts, images formed in infrared photography in night vision cameras, and in many other imaging devices like telescopes, etc. The main reason for dull images in case of X-ray imaging is the small number of X-rays that transmit through the anatomic object and form the image. The main purpose of an image intensifier is to convert the small number of X-ray photons coming through the anatomic object into many photons of ***visible light***. The intensity of the visible light photons delivered by the image intensifier is sufficiently large to make a strong and bright image. In short, an image intensifier converts X-ray photons into photons of visible light and also multiplies the number of visible light photons many fold over the number of X-ray photons. However, the operation of image intensification is carried out in several steps.

In an image intensifier (Fig. 5.2), X-rays coming out through the anatomic object are made to enter the image intensifier tube through the input window typically made of low-absorption low scattering material like aluminum or titanium. The X-ray beam then falls on a fluorescent input screen (also called the entrance phosphor), made of sodium activated cesium iodide CsI(Na), emitting a large number of visible light photons. It is the property of the entrance phosphor (activated cesium iodide) that converts X-ray photons into photons of visible light. The input fluorescent screen is followed by a photocathode. Visible light photons hit the photocathode and eject electrons from the photocathode. The electrons are accelerated towards the output fluorescent screen by an electric field produced between the photocathode and anode, as shown in the figure. Focusing electrodes are provided in the system to minimize distortions. The accelerated electrons finally hit the output phosphor made of silver activated zinc cadmium sulfide (ZnCdS(Ag)) creating a very large number of visible light photons, which are finally used to make a high-quality image of the anatomic object. The final visible light photons are taken out of the glass window and may be used for making a strong image. It may be noted that the final image in the output of the intensifier is smaller as compared to the image (X-rays) at its input. As such the image intensifier on one hand increased the number of visible light photons as compared to the X-ray photons and it also decreased the size of the image. Both these factors help in improving the image strength and quality. It may be recalled that exposing the photographic film to the visible light photons delivered by the image intensifier output changes the chemical structure of the film, but the image is still not visible. The exposed film needs to be chemically treated for three processes; ***fixing***, ***developing*** and ***drying*** before the final image become visible to the naked eye, even in bright background. It needs a dark room to carry out these processes. The above-mentioned three processes are part of conventional black–white photography, and therefore, no further discussion on these has been included. The image development

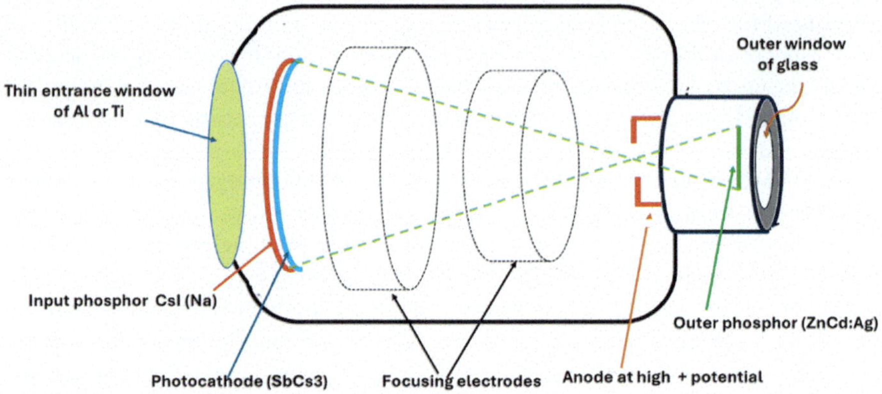

Fig. 5.2 Sketch of an image intensifier

process is time consuming, and after each exposure, a new film cassette is required for the next exposure.

Computed Radiography (CR): Computed radiography is the digital replacement of conventional X-ray film imaging technique and offers enormous advantages for inspection tasks; further, the use of consumables like photographic films is virtually eliminated. It drastically reduces the time consumed in processing the exposed film (developing, fixing and drying) in X-ray film radiography. Moreover, in CR the plates used to develop the image are reusable; the need for a dark room and chemicals for treating the film are eliminated. This reduces the cost and the exposure and processing times. The radiographic image may be evaluated by using the appropriate computer software. Since the data may be recorded in digital form, it becomes easy to exchange the data between different labs/physicians and may be stored in digital data libraries for future reference.

In CR, the X-rays emerging out through the anatomic object are made to fall on a CR panel. The CR panel is typically made by coating a plastic sheet with ***photostimulable phosphor materials*** like ($BaFBR:Eu^{2+}$). The property of the photostimulable phosphor material is that it absorbs the X-ray energy and releases ***clusters of electrons*** at the location where the X-ray is absorbed. The X-ray energy at these locations is stored in the form of ***Photo stimulated Luminescence Complexes***, the number of these complexes being proportional to the X-ray flux or intensity. As such, a two-dimensional lateral image of the anatomic object is formed on the CR panel by the electron clusters released from X-ray absorption (see Fig. 5.3a); obviously, this lateral image is not visible. This electron image is quite stable and remains embedded on the panel for a considerable period of time. To make a digital image from the electron image, the CR panel is illuminated by a strong and focused ***Laser beam*** row wise, one row after the other, till the complete 2-D panel is scanned by laser beam. Laser light hitting the electron complexes excites them, and while coming back to the ground state, these excited electrons/complexes emit visible light which is made to fall on a photomultiplier tube. The photomultiplier tube is like a transducer that

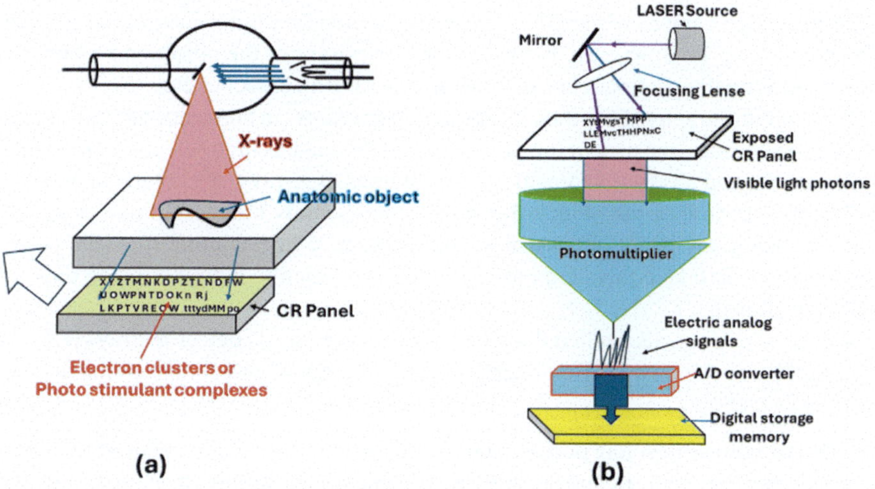

Fig. 5.3 **a** Formation of electron image on CR panel; **b** storing of digital image in computer memory

converts visible light into electronic signals. Thus, a lateral 2-D analog image of the object in terms of electronic signals is generated in the output of the photomultiplier tube as shown in Fig. 5.3b. The image made of electrical signals is fed to an analog-to-digital converter (A/D) to get the digital image. Digital images may be further analyzed using appropriate software or may be stored/shared with other labs or physicians. Any physician/technician may convert the digital image back into analog 2-D lateral image using a D/A converter.

To rub off all electron structures left on the CR panel after its use, the complete panel is again illuminated by a laser beam of an appropriate frequency. This brings back all electrons in the panel to their original place and energy states. This refreshes this panel, and the same panel may be used again for generating electron footprints of some other object.

Digital Radiography: Digital radiography (DR) is the latest and the fastest online X-ray imaging technique. It may be recalled that in computed radiography (CR), the X-rays transmitted through the anatomic object first create a lateral image, which is made up of electron clusters (or complexes) on the CR panel. The exposed CR panel is then taken to the reader assembly where it is scanned by a laser beam resulting in the emission of visible light photons that are converted into electrical signals using a photomultiplier. The electrical signals are then digitized and stored or used for creating the final image. In the case of the DR, however, the remnant X-rays instantly produce visible light photons which get converted into analog electronic signals. Analog electronic signals are immediately fed to a computer which digitizes them and generates the image almost instantly; the physician/technician can see the image on the screen seamlessly (instantly) as the object is put under the X-ray exposure. The CR technique may be compared to the process of taking photos by an old film-loaded camera while DR is like using a modern digital camera, no film, no film processing, no time delay between X-ray exposure of the object and seeing the image, no refreshing of panel or re-loading a fresh film cassette, etc.

Digital radiography (DR) uses ***X-ray sensitive plates*** (in conventional film screen and CR techniques one uses at the first stage a plate that has high sensitivity to visible light photons and a very low sensitivity for photons of X-rays) to capture data during the X-ray exposure of the object; the data is immediately transferred to a computer (without using any panel or cassette). The remnant X-rays coming from the object are converted into an equivalent electric charge and then to a digital image through a detector sensor.

Detector sensors, also called flat panel detectors or digital detector arrays (DDAs) used in DR, work on two different approaches, namely (i) direct conversion and (ii) indirect conversion.

(i) **Direct conversion flat panel detector** consists of two plain rectangular layers, one over the other; the upper layer is of some ***photoconductor*** material like the amorphous Selenium (amo. Se) or Cadmium telluride (Cd-Te). Photoconductive materials are materials that release electric charge when they absorb electromagnetic radiation or photons of gamma rays, X-rays, visible light, etc. This layer sits over a second plate having a matrix of ***multi microelectrodes***. Microelectrodes are very small electric connecting points. A microelectrode matrix is formed when many microelectrodes are uniformly distributed in a given area. Remnant X-ray photons (X-ray photons coming out from the anatomic object without being absorbed) coming through the anatomic object fall on the photoconductor layer that converts photons into electric charge, the liberated charge being proportional to the intensity/flux of the X-ray photons. The charge released by X-ray photons at different locations on the plate is picked up by the microelectrode located there. Each microelectrode transfers its charge to a ***thin-film transistor*** (TFT) network. A thin-film transistor (a very small electronic device) with its associated electronics converts the charge into an electronic pulse of the size proportional to the amount of charge which in turn is proportional to the intensity of the X-ray photons. These electronic pulses are amplified and digitized by the electronic network. The digital image data may be displayed on the screen in real time and/or may be stored.

Figure 5.4 shows the layout of a direct conversion system, the three 2-D sheets containing the photo conducting material, matrix of microelectrodes and the matrix of thin-film transistor networks are shown separated in the figure; however, in the actual system they are all one over the other. The TFT system amplifies the charge signals, converts them into electronic pulses and digitizes. This generates a 2-D digital dataset for the distribution of different shades of gray color in the X-ray shadow of the anatomic object.

(ii) **Indirect conversion flat panel detector** system also have two layers, top layer of some scintillator material (like CsI or NaI) which converts the remnant X-ray photons into photons of visible light of intensity proportional to the intensity

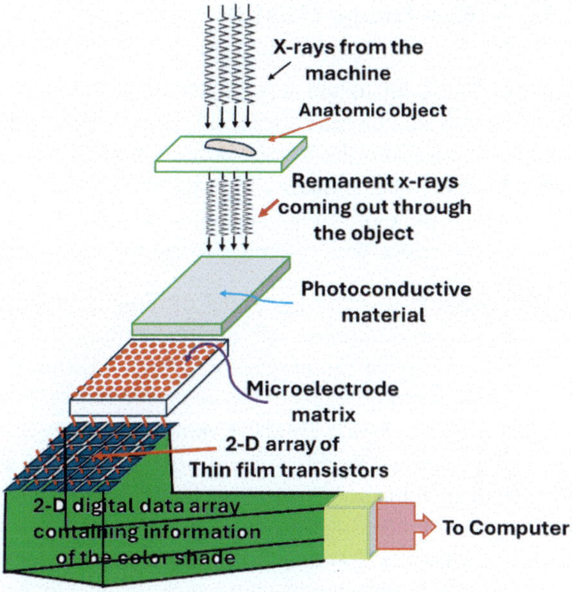

Fig. 5.4 Layout of a direct conversion flat panel detector system

of X-rays falling at that location. The second layer is a photo diode matrix of amorphous silicon to subsequently convert the visible light into charge. The charge is being proportional to the intensity/flux of the incident X-ray photons and inversely proportional to the thickness of the anatomic object immediately above. The outputs from different photodiodes of the matrix are fed to individual TFT networks which convert the charge information into digitized electric pulses, similar to the direct conversion approach.

Scintillating materials have the property that they emit a flash of visible light each time some X-rays (or charged particles) hit them. The intensity of the emitted visible light flash is proportional to the intensity of the incident X-ray beam. A photo diode is an electronic device which generates a charge pulse when illuminated with a visible light flash, and the size of the charge pulse is being proportional to the intensity of the incident visible light flash.

Figure 5.4 may also be used to show the layout of the indirect approach if one replaces the top layer of photoconductive material by a layer of scintillating material and the second layer (of microelectrodes) by a matrix of photodiodes. The output of each photodiode is then connected to the individual network of TFT transistors which processes the analog charge information and digitizes it for further applications.

5.5 X-Ray Image Quality

Quality of an X-ray image may be defined as the attribute of the image that influences the physician's/technician's level of certainty to visually perceive the required diagnostic features from the image. Although all X-ray images, obtained by whatever method, need to be of quality, but quality analysis is more appropriate for CR and DR imaging, the more frequent methods used at present. The quality analysis of the image may be done with reference to the following attributes of the image (Fig. 5.5).

(i) **Contrast:** It is the fractional difference in the intensity or the brightness of the region of interest and the surrounding areas in the image. Natural contrast in an X-ray image is generated by the difference in absorption of X-rays by different components of the anatomic object; bones absorb the largest fraction of incident X-rays (highest value of average absorption coefficient) followed by tissues and the least absorption takes places in regions that have air (lungs). Radiographic contrast is directly proportional to the density, thickness and the atomic number of constituent elements involved in the typical part of the object. X-ray energy that may be controlled by the anode voltage plays an important role in determining the contrast factor. The higher the energy of X-ray photons, less is their absorption in any part. Therefore, an X-ray image taken with low-energy X-rays shows better contrast as compared to the image of the same object taken by X-rays of higher energy. It is possible to enhance contrast of the desired organ or part of the image by putting **contrast enhancing dyes/chemicals** in the system before taking the image. For example, Barium salts are used to increase the contrast of the digestive tract, and iodine salts may be used for better radiography of gastrointestinal tract. In general contrast building agents, called dyes, enhance the absorption of X-rays in the specific part. Dyes get readily absorbed in the specific part or tract they are targeted.

(ii) **Spatial Resolution**: It is a measure of the ability of the imaging system to distinguish the adjacent structures separate from each other. It is observed that if one takes an X-ray image of a fine slit (a very narrow rectangular cut in a lead sheet), the edges in the X-ray image are not sharp, and the image appears blurred. Spatial resolution is a measure of the blur in an image and may be quantified in terms of the width of the blurred image at half of its maximum

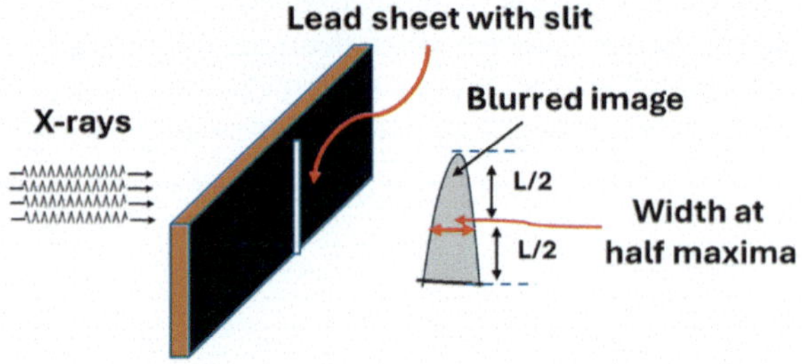

Fig. 5.5 Effect of spatial resolution in X-ray image

height. It is a property of the complete imaging system that generally does not depend on the energy (anode voltage), the intensity (filament current) of the X-ray machine and the time of exposure. Focal spot size, the amount of motion of the object during the X-ray exposure, type and size of the scintillator, if used in obtaining the image, energy of electrons in clusters formed in CR technique, etc., are the factors that affect spatial resolution.

(iii) **Dynamic Range:** It is the range of X-ray intensities which produce a good image of some object by a given set of detecting systems. It is known that the intensity of the X-ray beam increases with the increase of the filament current of the X-ray tube. Suppose one carries out an experiment in which the same object is X-rayed for different filament currents, starting say from I_0 to a maximum current value of I_f mA (milli Ampere). In such an experiment, it will be observed that X-ray images of good quality are obtained only over a span of currents say from I_x to I_y mA corresponding to X-ray intensities in the range of I_a to I_b. The range of intensity I_a to I_b is called the dynamic range. A large value of dynamic range is considered a good quality of the imaging system. Essentially, dynamic range is a property of the detector system used in radiography.

(iv) **Noise:** The actual image of an object in an X-ray radiograph is produced by the variation in the absorption of X-rays by different parts/constituents of the object. However, often one finds random or structured variations in the X-ray intensity within the image of an object that does not correspond to the variation in the absorption. These random or structured variations in X-ray intensity unrelated to the differential absorption are noise. Noise is undesirable and may arise due to several reasons including the presence of scattered X-rays. A significant contribution to noise comes from ***quantum noise or quantum mottle***, which is due to fluctuations in the number of photons reaching the detector from point to point. Quantum noise decreases with the increase in the number of X-ray photons, that is with the exposure time. Noise makes interpretation of X-ray images more difficult.

(v) **Artifacts**: Artifacts are the structures seen in an X-ray image that are not present in reality; they appear mostly from objects outside the patient. For example, X-ray shadows of patients' clothing, external cardiac monitor leads, body parts of the carer, etc., show up in the X-ray image. Artifacts obscure or distort the real image.

(vi) **Signal-to-Noise Ratio**: Signal-to-noise ratio (SNR) is a parameter that reflects the combined effect of contrast, spatial resolution and noise. A higher value of SNR for an X-ray image allows recognition of smaller and lower contrast structures.

Chapter 6
Computed Tomography (CT) or CAT Scan

Abstract Tomography is the science of studying the properties of a solid body by dividing the body into thin slices and recording the characteristics of these slices. Once the characteristics of these individual slices are studied, it is then possible to generate a detailed picture of the complete body, synthesizing the individual slice pictures using a computer. It is like cutting a loaf of bread into toasts and by observing the quality or nature of individual toast to conclude about the bread loaf. Technique of computed tomography, its advantages, etc., are discussed in this chapter.

6.1 Introduction

Computed tomography (CT) or computed axial tomography (CAT) uses a combination of X-ray imaging and a powerful computer to create images of the internal body parts; bones, organs and other tissues to provide more detailed information as compared to a normal X-ray image. The word 'tomography' refers to a series of detailed pictures of areas/cross sections inside a body. It may also be referred to as a method of producing a three-dimensional image of the internal structures of an object, such as the loaf of a bread, by taking images of different cross sections of the object (for example, the slices of the bread loaf) and compiling them. However, in the present context, it includes the human body and the images taken by X-rays with the help of a computer. In CT scanning, a narrow and focused X-ray beam is aimed at the desired body part of a patient and the source of X-ray is quickly rotated around the body part. The residual X-ray beam that passed through the body part (unabsorbed) is detected by a suitable detector array (placed directly opposite to the X-ray source on the other side of the body part) and is converted into electrical signals which are digitized and analyzed by a computer to generate the X-ray image. X-ray images of the same thin slice of the body part taken from several positions of the X-ray source in its circular motion are synthesized by the computer to produce the detailed X-ray image of a thin slice of the body part. In this way, images of the successive slices of the body part may be taken by moving it (the body part by a small distance) after each complete rotation of the X-ray source. The successive image slices may then

R. Prasad, *Understanding Medical Devices*, Series in BioEngineering,
https://doi.org/10.1007/978-3-031-88630-0_6

be used to generate the complete three-dimensional image of the body part. Since the number of images generated in the process is very large, they are analyzed by a computer.

The first CT scanner was invented by an electrical engineer, Sir Godfrey Hounsfield in 1972. A physicist, Allan McLeod Cormack also developed a similar system around the same time, and both Housenfield and Cormack shared the Nobel Prize in Physiology/Medicine for their invention in 1979.

It may be realized that CT scan contains two-dimensional X-ray images that represent a three-dimensional object, and therefore, interpreting these images is a difficult task.

In CT scanning, a narrow beam of X-rays is aimed at the patient and the beam is quickly rotated around the body of the patient. The X-rays transmitted through the body are recorded by a detector placed directly opposite to the X-ray source and rotates with the source. Rotating X-ray beam enters a given part of the patient's body successively at different angles, producing X-ray images of the targeted body part taken from different directions or angles. The thickness at different angles and the type of substructures in the targeted part of the body, as seen by the X-rays passing through at different angles/directions, are reflected in these X-ray images. X-ray images are taken at fixed time intervals that may be adjusted to the desired value. Digitized X-ray images taken in one complete rotation of the X-ray source are processed by the inbuilt minicomputer of the system that generates a two-dimensional image section called '***slice***'. The bed or bench on which the patient lies is then moved by a small distance (of the order of few cm) and the X-ray source takes another round, generating another slice of X-ray image of the successive section of the patient's body. In this way, the whole body of the patient may be scanned generating many X-ray image slices. Each X-ray image slice may be compared to an individual toast of a bread loaf. By piling the successive image slices one after the other, a three-dimensional image of the whole body may be made.

6.2 CT Scan Machine

A CT scan system or machine has the following main components:

(i) A power supply unit (ii) A control console (iii) Gantry or Scanning unit and (iv) A moveable patient table. A rough sketch of the gantry and the patient table is shown in Fig. 6.1a. The power supply unit and the control console are separate from the granary and the patient table, the heart of the system, are not shown in this figure. The power supply unit supplies the electric power to the X-ray generating tube, detector array and to various electronic and electrical systems housed in the gantry. Since the X-ray tube and the detector array, housed in the gantry, are made to rotate in a circular path within the gantry, the power supply system can supply the required power to these moving systems without any interruption. The control console is a unit from where operation commands

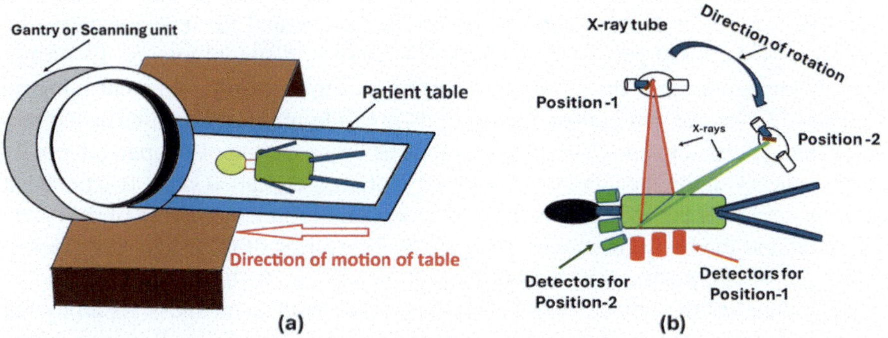

Fig. 6.1 **a** Rough sketch of a CT scan system, **b** X-ray image recording for two positions of the X-ray tube and detector array

are issued to operate the system. Person at the control may alter the speed of rotation of systems within the gantry, may change the vertical tilt of the gantry, may move the patient table to bring the desired body part under the gantry for X-ray scan, may control the exposure time, etc.

Gantry is a doughnut-shaped circular ring-like structure held in vertical plane. A horizontal flat table on which the patient (or any other object under investigation) may lie down with back on the bench or in the inverted position is also provided. The patient table may be moved into the hollow circular space at the center of the gantry, remotely from the control console.

In basic CT scanners, the gantry contains an X-ray tube that provides a well-focused and defined X-ray beam. There are beam defining slit systems for this purpose. The X-ray beam comes out of the gantry through an aperture into the hollow space and may hit the desired body part of the patient lying on the table. Diametrically opposite to the X-ray tube, on the other side across the patient table, there is another aperture in the gantry wall. X -rays after passing through the desired body part of the patient enter the gantry through the aperture and fall on the array of detectors. Detectors convert the X-ray energy into electric pulses that are digitized by the A/D converters housed in the gantry. Digitized X-ray image data is further processed by a powerful minicomputer, also held in the gantry.

As shown in Fig. 6.1b in their rotatory motion around the target body part, X-rays pass through the body part at different angles probing the part along the incident plane. Thus, each X-ray image recorded at different positions of the X-ray tube contains information of a cross section of the target body part along the direction of the incident X-ray plane. A large number of X-ray images of the same slice of the body part are thus recorded in one complete rotation of the X-ray tube. The inbuilt computer analyzes these different images taken in a complete rotation and generates a final detailed image. Thus, one detailed image of the cross-sectional area of the body part is generated for each revolution of the X-ray source. After the completion of each rotation of the X-ray tube, the patient table is moved forward by a fixed distance, say ‘d’. The machine operator may set the magnitude of the distance ‘d’ remotely from his console. In the next rotation of the X-ray tube and the detector array, the next slice of the body part is scanned by X-rays from different angles. To further illustrate the working of the scanning mechanism, Fig. 6.2 shows a patent lying on the table. The central part of the patient's body in this figure is represented by a three-dimensional cuboid to simplify the drawing. The cross section of the patient's body lying vertically under the gantry (rectangular in this case), called ***slice***, is scanned at different angles by the rotating beam of X-rays in one complete rotation of the X-ray source around the body, as shown in Fig. 6.3a. Two slices of the body part scanned by X-rays for two positions of the body are shown in Fig. 6.2b.

Each slice has the object image in the x–y plane with a thickness Δz along the axis of the gantry. The thickness Δz is governed by the spread or collimation of the

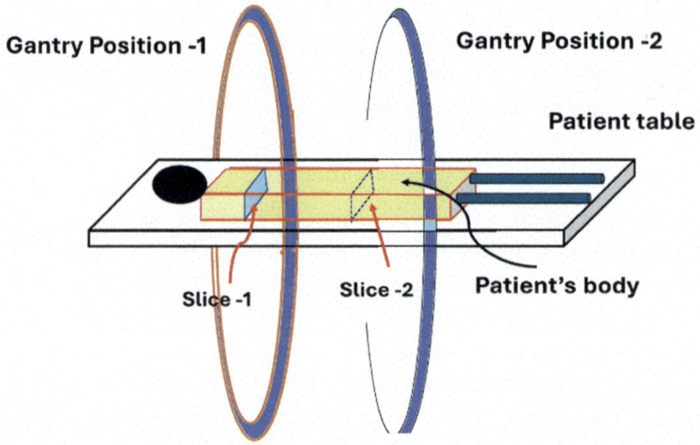

Fig. 6.2 Figure shows the two slices of the body part scanned by the X-ray beam for two positions of the patient's table

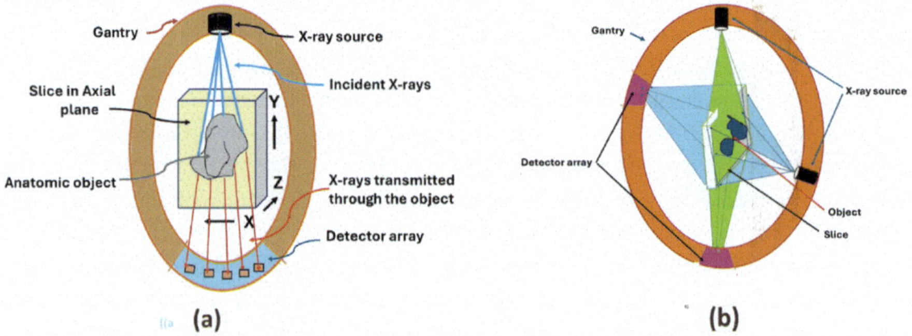

Fig. 6.3 **a** Axial slice of a CT scan image. **b** In orbital motion of the X-ray source and the detection array around the object, several images of the same object are taken at different angles; the situation for two positions of the X-ray source is shown in this figure

X-ray beam in the Z-direction. The ratio d/Δz is called the ***pitch*** of the machine. In the case, the distance 'd' moved by the patient table after each rotation is equal to Δz; the pitch becomes 1 (unity) and successive slices contain continuous images of the object without any gap.

An image slice may be divided into several very small cuboids each of volume $\Delta V = \Delta x \cdot \Delta y \cdot \Delta z$; as shown in Fig. 6.4. These volume elements are called ***voxels***. However, the final image of a slice made by the computer is projected either on a film or a LED screen, both of which are only two dimensional. As such image information contained in each voxel of the slice is embedded in a corresponding elementary area called ***pixel*** on the film or screen, as shown in Fig. 6.4. Resolution of an X-ray image depends on the number of pixels in the image, which in turn depends on the pixel size.

In principle if the image slices are piled one after the other a three-dimensional image of the target body may be generated. Since the number of images taken in the scanning process is very large, it is not possible to generate the three-dimensional profile manually. A fast computer with analyzing software associated with the system processes the image data and prints it in a specific sequence on a two-dimensional film/paper sheet or displays it on the screen. Radiologists are trained to read and decipher the features of these two-dimensional image data. It is customary to present the projection of the 3-D X-ray image of the object in one of the three, mutually perpendicular planes called the ***axial plane***, the ***coronal plane*** and the ***Sagittarian plane***. These planes are shown in Fig. 6.5.

While reading an X-ray image is a specialized job, however some basic image features may be easily understood. In most cases, each X-ray image is made up of areas with different shades of black and gray colors, corresponding to the intensity

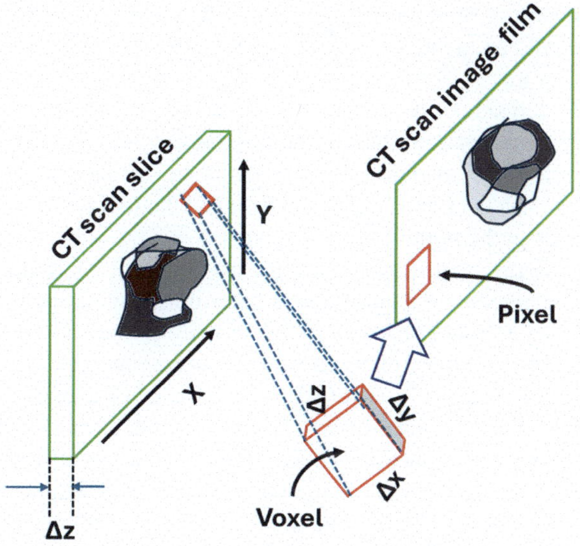

Fig. 6.4 Elementary volume element, called voxel, of the image slice is represented by an area element of area $\Delta A = \Delta x \cdot \Delta y$, called pixel

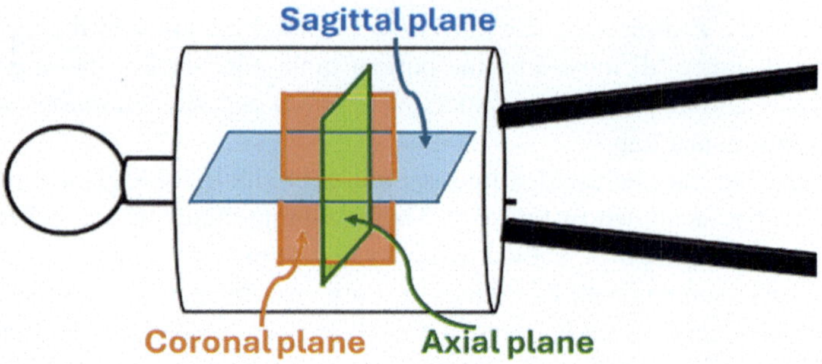

Fig. 6.5 Three important planes along which the two-dimensional image of a three-dimensional object may be projected

of the X-rays allowed to pass through different parts of the target body. It may be remembered that an area of dark black shade in the scan image indicates that the corresponding body part is quite transparent to X-rays, has not absorbed much of it and therefore of low density (fat or other low-density tissues). A very white area of X-ray image tells that the corresponding body part is of high-density matter (like bones) that has absorbed considerable intensity of the incident X-rays. In special cases one may also get colored X-ray images.

6.2.1 Reading a CT Scan Image

Before going into the details of an image, it is required to hold the X-ray image film in proper orientation. The words **R** and **L** on the film tell which side of the film should be facing the reader and the direction of the top of the film. Looking at a CT scan film is like looking into a mirror; the right side of the patient's body will be on the left side of the reader. Letters R and L on the film specify the right and left sides of the patient, not that of the film paper.

It is a convention to show the X-ray images of the anterior or front part of the body on the top of the X-ray film and the images of the posterior or back part of the body at the bottom of the film.

X-ray imaging is based on the principle of X-ray attenuation (or absorption). The amount of attenuation is determined by the density and atomic number of the imaged body part/tissue. High density tissues like bones absorb large intensity of X-rays and transmit a reduced amount on the other side to the detector; hence, they appear brighter or whitish in the image. Low-density tissues like lungs, fat, etc., absorb X-ray radiation to a lesser degree and transmit a larger intensity of X-rays to the detector, appearing darker in the image. Pockets where air is trapped in the body have the least absorption of X-rays and appear dark black in image film. Different shades of gray in the image correspond to different types of fluids including blood in the imaged body part.

Table 6.1 Hounsfield values for some shades of black color

Material	Hounsfield value	Color in image film
Air	− 1000	Black
Fat	− 40 to − 100	Dark gray
Water	0	Gray
Blood	+ 70	Light gray
Bone	+ 400 to + 1000	White

In radiology the attenuation power of different tissues is quantified by the ***Hounsfield scale*** denoted by (HU). Hounsfield values for some typical substances are given below (Table 6.1).

Digitized electric signals corresponding to transmitted X-rays may be converted into Hounsfield values, and each Hounsfield value may be assigned a different color to generate a colored image of the slice. This conversion process is possible only with the help of a fast computer. Often, a contrast enhancing dye is also used to show some structures more clearly. For example, if a 3-D image of the digestive system (abdomen) is required, the patient may have to drink a barium meal. The barium appears white on the scan image as it travels through the digestive system.

As mentioned, in CT scanning a 3-D image of the target body (object), in digital format, is generated by the computer and the digital image data is stored in computer memory in a specific sequence. The computer, according to the instructions given by the algorithm, may project the image of the object in one of the three planes, axial, coronal or sagittal. Typical CT scan images of human face in three planes are shown in Fig. 6.6,

In modern CT scan machines, there is not one but many sources of X-rays and associated detector arrays in gantry. This reduces the recording time and improves the resolution of the image.

X-rays are high-energy electromagnetic radiations, which may ionize the matter through which they pass. Ionization is the process in which electrons attached to different atoms of matter get detached from the atom. Ionization electrons released

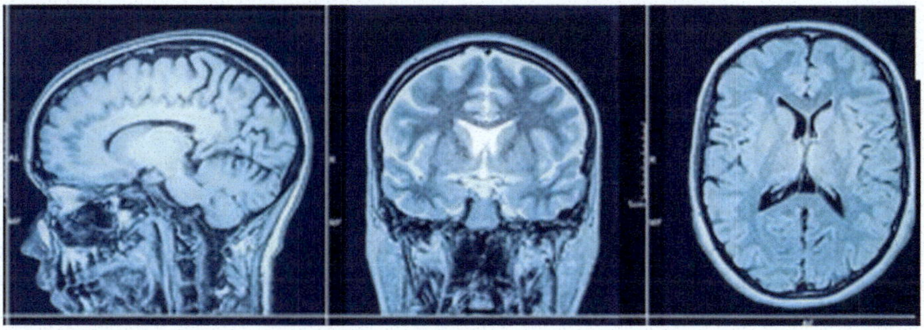

Fig. 6.6 Typical CT scan image of face in three planes

by the passage of X-rays are dangerous for living cells as they have sufficient energy that may destroy the cell. Therefore, repeated CT scanning is harmful for the living objects. This is a big disadvantage of any type of X-ray scanning.

Chapter 7
Magnetic Resonance Imaging (MRI)

Abstract Magnetic resonance imaging is a beautiful example of how the magnetic properties of the human body may be used to generate images of body parts that have abundant hydrogen atoms surrounded in a favorable environment, where they may readily change their orientation. Each proton, the nucleus of a hydrogen atom, behaves like a spinning top with a magnetic field. When this proton (or hydrogen atom) is embedded in a medium like water, body tissues, ligaments, etc., it may easily change its orientation by emitting a special electronic signal that carries information about the nature of the surrounding medium (tissue, flesh, etc.). If the hydrogen atom or the proton is surrounded by teeth bone and other similar media, it takes much longer time to change its orientation and therefore does not emit a useful electronic signal. Detail of magnetic resonance technique is discussed in this chapter.

7.1 Introduction

Magnetic resonance imaging (MRI) is a method of generating images of body parts using magnetic properties of the body. The magnetic properties of the body are derived from the distribution of hydrogen atoms in different body parts. Hydrogen atoms in the form of water molecules are abundantly distributed in fat, kidney, gray and white matter of brain, muscle, liver, tendon, proteins, blood and other fluids in the body; but hydrogen atoms are not so abundant in structures like bones and teeth as water contents in them is almost absent. We concentrate on water content because in a water molecule the relative percentage of hydrogen atoms and hence of protons is much higher as compared to the percentage of protons in any other heavier molecule. Therefore, MRI procedure is not very effective in taking images of teeth and bones. This method of imaging requires, apart from protons in the body, an external source of strong magnetic field, sources of Radio frequency (RF) waves of different frequencies and a powerful computer. Radio waves (RF) are electromagnetic waves of very low frequencies, and they carry a very small amount of energy as compared to X-rays; therefore, unlike the X-rays, radio waves do not ionize the medium through which

R. Prasad, *Understanding Medical Devices*, Series in BioEngineering,
https://doi.org/10.1007/978-3-031-88630-0_7

they pass. MRI imaging is not associated with any health hazard in contrast to the X-ray imaging and the CAT scan both of which use ionizing X-rays.

7.2 The Proton

To understand the role of a proton in MRI, it is required to know more about the proton. Proton is a fundamental particle and is a constituent of all matter. As we know, materials are made up of atoms of different elements that join to make a molecule of the given material; a molecule of a given material exhibits all physical and chemical properties of that material and is the smallest entity that may stay freely. When a molecule of some material is broken, it dissociates into atoms of specific elements. Thus, atoms of different elements are the building blocks of all materials found on the earth. There are almost 80 stable elements and more than 40 unstable (also called radioactive) elements that have been identified so far. It is also observed that some atoms have their isotopes that differ with each other in terms of their mass and stability. Most of the atoms are almost spherical in shape, and atoms of different elements have different mass and size (radius). For example, the lightest atom is the atom of hydrogen that has a mass of $\approx 1.6735 \times 10^{-27}$ kg and a radius of $\approx 37.0 \times 10^{-12}$ m, while the heaviest stable atom is that of Uranium-238 that has a mass of $\approx 395.3 \times 10^{-27}$ kg and radius of $\approx 7400 \times 10^{-12}$ m. Each atom has a nucleus at its center where the total positive charge and almost all mass of the atom reside; atomic nucleus is also a tiny spherical volume at the center of the atom, for example, the radius of the nucleus of hydrogen atom is about 1.2×10^{-15} m while that of uranium atom is about 7.4×10^{-15} m. In atoms, the negatively charged electrons rotate around the positively charged nucleus, like planets rotate around the Sun. In atomic/microscopic world charge is measured in units of the charge on an electron which is denoted by the letter '**e**'. +1e of charge is equal to $+ 1.60 \times 10^{-19}$ Coulomb; here Coulomb (denoted by C) is the unit of charge in the metric system.

Figure 7.1 shows how a piece of matter may be looked at as an aggregate of atoms which in turn is made up of nucleus and electrons. Atomic nuclei are made up of two particles: the proton and the neutron. Each proton carries +1e unit of (positive) charge, while the neutron is chargeless. The nucleus of hydrogen atom has only one proton and no neutrons, while the nucleus of Uranium-238 atom has 92 protons and 146 neutrons and the nucleus of Carbon-12 has 6 protons and 6 neutrons. It may be observed that the relative weight of protons in a water molecule is much higher than in a molecule made up of other heavier atoms.

Properties of a proton may be summarized as follows:

Mass $= 1.67 \times 10^{-27}$ kg; radius $= 0.85 \times 10^{-15}$ m; charge $= +1\text{e} = 1.60 \times 10^{-19}$ C; spin $= 1/2\ \hbar$, behaves like a tiny bar magnet.

Mass, radius and charge are observables that are well known; however, ***Spin*** is a new property that many microscopic particles possess. Though spin is a quantum mechanical concept, in very simple terms it may be associated with some kind of a rotation of the particle about an axis passing through its center, like that of the

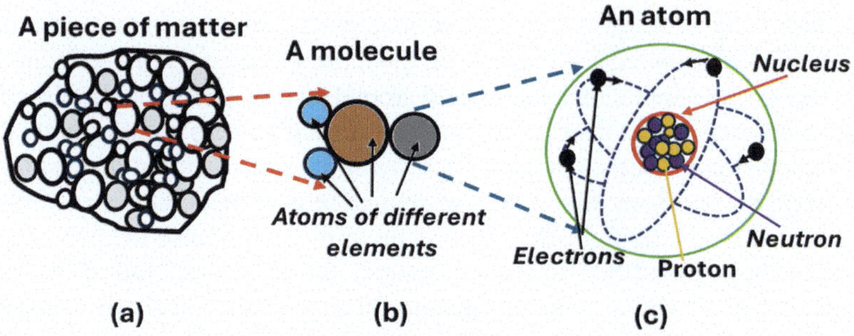

Fig. 7.1 **a** A piece of matter **b** a molecule and **c** an atom

rotation of a toy top or the rotation of the earth on its axis. Spin which is generally denoted by the letter **S** is measured in a unit called ***'rationalized Plank's constant'*** denoted by the symbol $\hbar$. The spin of a proton is denoted as $S = 1/2\ \hbar$. In order to specify the direction of rotation of the proton with respect to some direction in space, say Z-axis, it is assumed that the direction of spin is defined by the right-hand thumb rule shown in Fig. 7.2a.

Two protons with spins $+ 1/2\ \hbar$ and $- 1/2\ \hbar$ are shown in Fig. 7.2b. It may be shown that a spinning charged particle, like a proton, behaves like a tiny bar magnet, as indicated in Fig. 7.2c.

Each hydrogen atom has a proton and therefore, human body parts that have large contents of water molecules, like tissues, fat, tendon, gray matter, white matter, blood, other fluids, etc., carry many protons, each behaving like a tiny bar magnet. The magnetic property of these protons is used in developing images of anatomic objects using the MRI technique.

Relaxation Times T1 To understand the working of an MRI system, let us consider a typical body part that has considerable molecules of water or hydrogen atoms. Each hydrogen atom contains one proton, and therefore, the given body part has many

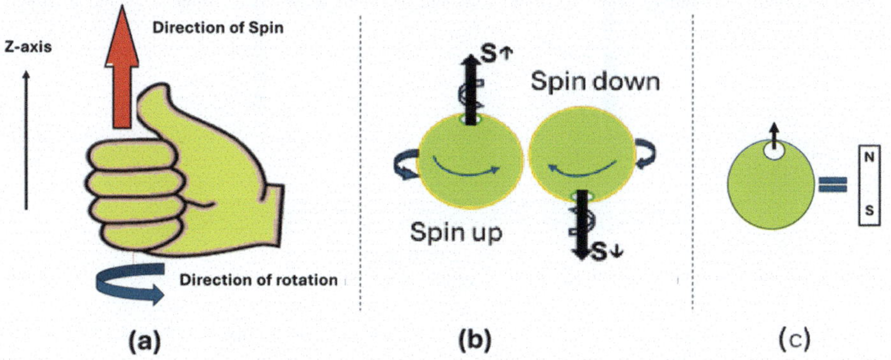

Fig. 7.2 **a** Right-hand rule to specify the direction of Spin, **b** two protons with $+ 1/2\ \hbar$ and $- 1/2\ \hbar$ spins, **c** a proton is like a tiny bar magnet

protons. Normally protons in a given anatomic part, contained in hydrogen atoms, have their spins distributed randomly in all the three space directions X, Y and Z; as a result, there is no resultant magnetic field around, and magnetic fields of randomly distributed protons cancel each other. So, we start with a zero equilibrium value of the magnetic field. It is shown in Fig. 7.3a. Now suppose an external strong magnetic field of strength, say B_0 in direction Z, is applied to the object. Under the influence of the external magnetic field B_0, most of the protons, because of their magnetic fields, align their spins (their tiny bar magnets) along the direction of the applied magnetic field B_0. This alignment of proton spins generates an **internal magnetic field** of maximum value, say M_0, parallel to the direction of the external field B_0.

An important aspect of the process of the proton spin alignment is that it is not instantaneous; all protons present in the anatomic object do not immediately align their spins in the direction of the external field B_0. It means that the internal magnetic field M grows slowly from an initial zero value to the final maximum value M_0. The process of spin alignment is called the ***Relaxation Process***. The growth curve for the internal magnetic field M is shown in Fig. 7.4.

The time in which the internal magnetic field grows to a value of 63% of the maximum value M_0 is defined as the ***relaxation time T1***. It is interesting to note that relaxation time T1 strongly depends on the environment around the protons, i.e., on the type of tissues that surrounds the protons. Relaxation time T1 may vary from a fraction of a second to several seconds. Table 7.1 shows the value of relaxation times for some tissue types.

The dependence of relaxation time T1 on the nature of the surrounding tissue may be understood since the efforts or work done by the proton in rotating or in changing the direction of its spin will obviously depend on the nature or the tissue

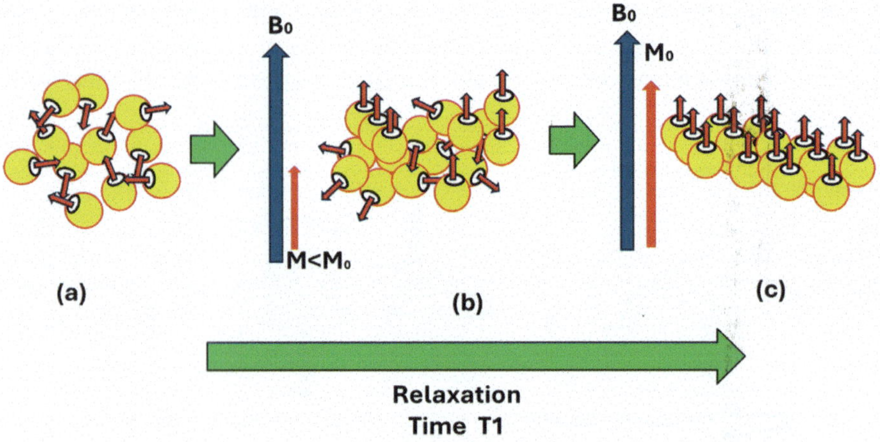

Fig. 7.3 **a** In an assembly of protons, their spin directions are randomly distributed, **b** under the influence of a strong external magnetic field B_0 proton spins start aligning along the direction of the external field and gradually an internal magnetic field starts developing, **c** in a short time most of the protons align their spins in the direction of external magnetic field B_0

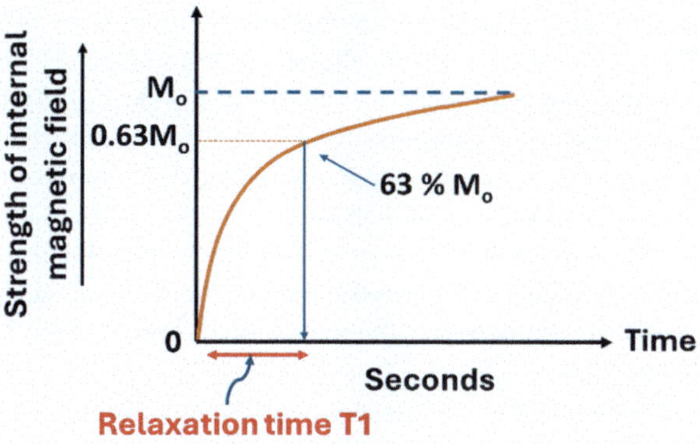

Fig. 7.4 Internal magnetic field grows with time after the application of the external magnetic field. The time in which the internal magnetic field reaches 63% of its maximum value M_0 is defined as the relaxation time T1

Table 7.1 T1 and T2 relaxation times for some materials

Tissue type	T1 in sec	T2 in sec
Ice	5.0	1×10^{-6}
Water	4.0	2.0
Muscle	0.9	0.09
Liver	0.50	0.04
Fat	0.25	0.07
Tendon	0.40	5×10^{-3}
Proteins	0.25	1×10^{-3}
Gray matter (brain)	1.5	9×10^{-2}

type surrounding the proton; some tissues may allow proton to turn easily while the other may need more energy and efforts. In body parts like teeth and bones, there are protons in different atoms and molecules, but the surrounding environment around these protons do not allow them to turn their spins easily, the proton spin relaxation time for such body parts run into tens of seconds. Such slow processes are not useful for MRI imaging.

7.3 Relaxation Time T2

Now if these aligned protons are bombarded for some time by radio waves (electromagnetic waves of low frequency and of large wavelength) of a ***specific frequency*** ν_0, then most of the aligned protons will absorb energy from the radio waves (RF waves) and will turn around by 90° to confine their spins in the X–Y plane. The process of absorption of energy by the aligned proton from the RF wave is called the ***Nuclear Magnetic Resonance***, from which the technique gets its name. The specific frequency ν_0 of the RF wave is called the ***resonance frequency***. The resonance frequency may be defined as the frequency of the RF wave from which the aligned protons may absorb energy and turn their spins by 90°. The value of the resonance frequency for a group of aligned protons depends ***only on the magnitude of the external magnetic field B_0*** (and does not depend on the nature of the surrounding tissue type). For example, if v_0 is the resonance frequency for a group of protons aligned in Z-direction by a magnetic field B_0, then for the same group of protons the resonance frequency will change to (say) v_1 if the external magnetic field is changed from B_0 to B_1. It will be shown later that by keeping different body sections at slightly different values of external magnetic fields, each section responding to a different resonance frequency v, the nature of tissues in different slices of the body may be determined.

Bombardment of protons aligned in Z-direction with RF waves of resonance frequency continues for some time so that most of the aligned protons absorb energy from the RF wave of resonance frequency and turn around to confine their spins in the X–Y plane, as shown in Fig. 7.5a and b. It may be realized that protons with spins in the X–Y plane have extra energy which they have absorbed from the RF wave of resonance frequency, and therefore, protons with spins in the X–Y plane are not in a stable state (it is a well-known principle of science which says that any system that has excess energy is unstable and tries to revert back to the lower energy stable state). Further, when proton spins are confined in the X–Y plane, there is no internal magnetic field in Z-direction, and there is only external field B_0 in the Z-direction. An internal magnetic field may develop in the X–Y plane for a short time that finally becomes zero when spins in the X–Y plane get randomized. After some time, the bombardment of protons by the resonance frequency RF waves is stopped. At the stop of the bombardment by RF wave, protons with their spins in the X–Y plane interact with each other through their magnetic fields and try to randomize the directions of their spins to attain the state of lower energy. Randomization of spins, again a relaxation process, is also not instantaneous; it takes some time, the time in which 63% of the spins get randomized is called the ***relaxation time T2***. It may be observed in Table 7.1, that T2 also has specific value for the given tissue type. As such if the value of both T1 and T2 for a given tissue is known, it is possible to identify the nature or type of the tissue. In very sophisticated and precision MRI systems, relaxation times T1 and T2 are measured, and the type of the tissue is determined by their measured values. However, direct measurement of T1 and T2 is quite involved and requires very costly time measuring systems. In normal MRI systems instead of measuring the two relaxation times, the intensity of the RF radiation emitted by the aligning protons is measured. The process is described below.

Once the randomization of spins is complete, the proton spins start aligning in the direction of the external magnetic field (field B_0 in Z-direction). Randomization of proton spins and their alignment in Z-direction are associated with the emission of extra energy in the form of radio (RF) waves. The RF radiation emitted during spin randomization and the spin alignment may be detected using RF receiver coils. The intensity of the emitted RF signal is closely related to both the relaxation times T1 and T2 and therefore is the identifier of the tissue type. In most of the MRI systems, a calibration experiment is generally carried out by the manufacturer, and the data of the calibration experiment is stored in the computer of the MRI system. In the

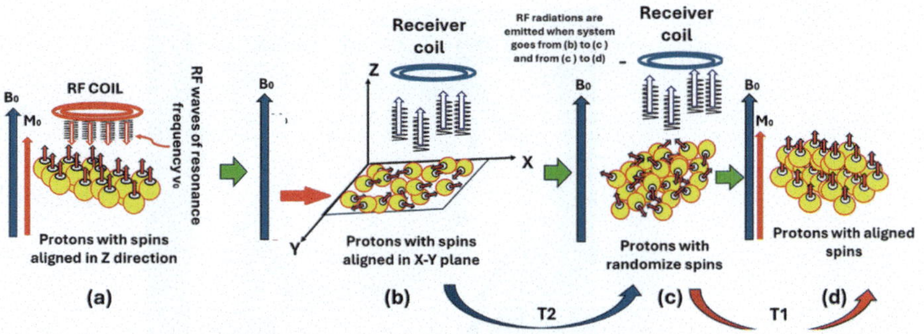

Fig. 7.5 **a** Protons with spins aligned in Z-direction are bombarded by RF waves of resonance frequency ν_0, **b** aligned protons absorb energy from RF waves and turn their spins in the X–Y plane. When bombardment by RF waves is stopped, protons randomize their spins emitting RF radiation. The process of spin randomization takes some time characterized by the relaxation time T2. **c** Once the spins get randomized there is no internal magnetic field, the protons again start to align their spins in the direction of the external magnetic field to generate the internal magnetic field with characteristic relaxation time T1

calibration experiment known tissue types are taken, one at a time, and protons of the known tissue sample are aligned by a strong magnetic field; then RF waves of resonance frequency are made to hit the sample. This results in turning the spins of the protons in the X–Y plane. The resonance RF wave is then switched off, and the intensity of the RF signal emitted by the aligning protons is recorded. RF signals emitted by aligning protons are not of the same intensity because of the statistical nature of the process. The inbuilt computer of the system calculates the mean value of the emitted RF signal intensity and correlates it with the tissue type. In this way, by taking samples of different tissue types, a calibration table between tissue type and average intensity of the emitted RF signal is generated and stored in the computer. This calibration table is used later to identify the type of tissue from the measured RF signal intensity in the case of a patent. It may be remembered that calibration and subsequent identification of tissue type from the intensity of the RF signal, etc., are all carried out by the computer. The final image is generated by assigning different shades of gray color to different intensities of the RF signal. It is customary to associate high (RF signal) intensity to white color, intermediate signal intensity to gray color and low signal intensity with black color in the MRI image. MRI techniques are often used to study metabolism of the brain. Figure 7.6 shows the MRI scan of the knee joint; muscles and tendons are very clearly seen in this figure.

Brain MRI images are often colored. The color scheme is designed so that areas with black and blue colors represent healthy brain metabolism, while areas represented by green, yellow and red colors indicate progressively worsening brain metabolism, as shown in Fig. 7.7.

MRI Scanners MRI scanning machines are available in several sizes, big machines for whole-body scanning, smaller versions for specific body part scanning. Let us consider a whole-body MRI scan system. In most cases, it is like a horizontal tunnel

Fig. 7.6 MRI scan of knee

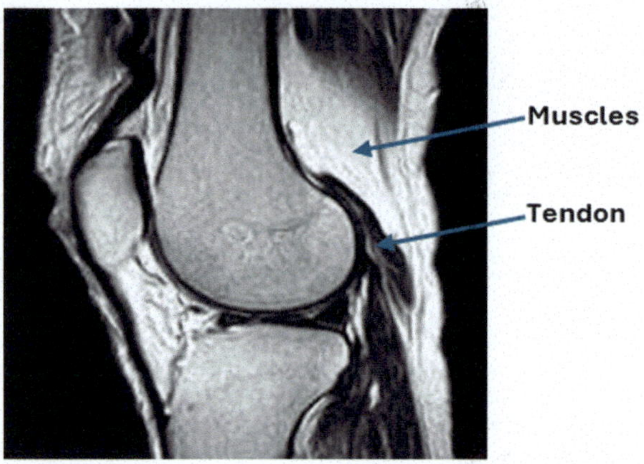

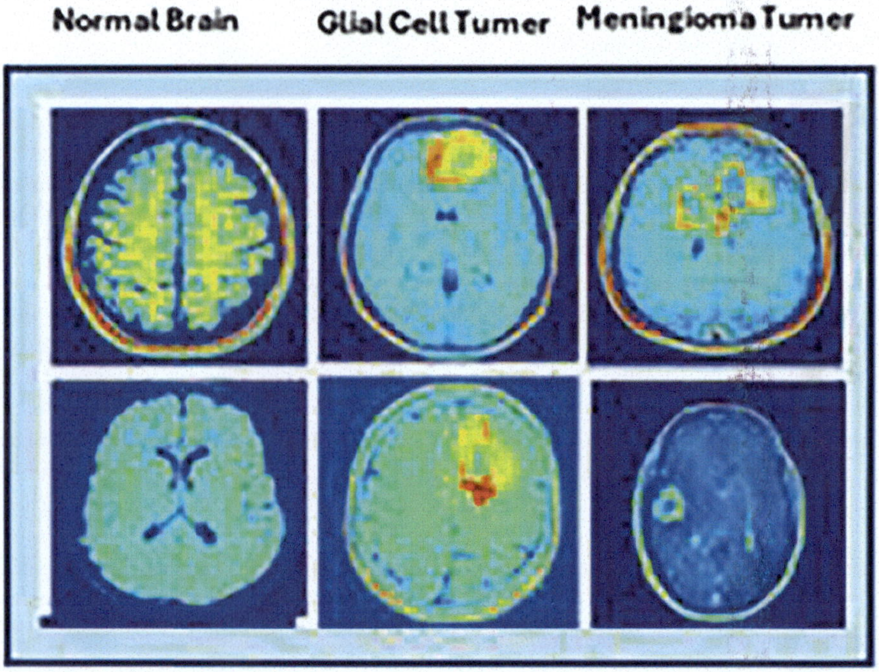

Fig. 7.7 MRI images showing brain metabolism

with a thick cylindrical cover. The patient on a movable table may be made to slide in the tunnel cavity and the tunnel gate is closed. The cylindrical tunnel cover carries the following basic structures; (i) ***A pair of large electromagnet coils*** that may produce a strong magnetic field B_0 from head to toe of the patient. The strong magnetic field aligns protons in different body parts of the patient, generating internal magnetic fields. (ii) ***Gradient electromagnet coils are*** pairs of small coils, say '*n*' in number, distributed all over the length of the patient's body at small distances. Each pair of gradient coils adds a small magnetic field to the large magnetic field B_0. The ith gradient coil adds a small magnetic field (ΔB_i) to the field B_0 on the part of the patient's body immediately under the gradient coil, thus increasing the strength of the magnetic field from B_0 to $B_i(= B_0 + \Delta B_i)$ in the body part immediately below the ith gradient coil. In this way, the body of the patent is divided into ***slices*** that have magnetic fields of different strengths $B_0, B_1, B_2, \ldots$. Since the value of the resonant RF wave that may be absorbed by protons depends on the strength of the magnetic field, protons in different slices of the body will absorb energy from RF waves of different resonance frequencies $\nu_0, \nu_1, \nu_2, \ldots$, etc. The idea is that when RF wave of a particular frequency say ν_2 is incident on the patent, only the aligned protons in the slice of the body that has the magnetic field of strength B_2 absorb energy and turn their spins in the X–Y plane; protons in other body parts (or slices) are not affected by the RF wave. This is how tissue structures of the successive vertical body cross sections (slices) may be studied on switching radio waves of frequencies $\nu_0, \nu_1, \nu_2, \ldots$, etc., one after the other. (iii) ***RF coils that may generate and radiate*** to the patient body radio waves of different resonance frequencies, $\nu_0, \nu_1, \nu_2, \ldots$, etc. These coils may generate and radiate out to the body of the patient RF waves of frequencies, $\nu_0, \nu_1, \nu_2, \ldots$, etc. (iv) ***Receiver RF coils*** that serve as antennas to receive signals emitted by protons undergoing the process of relaxation. Receiver coils feed RF signals received from the patient's body to a transducer that converts them in digital electric form and measures their intensity. Rough sketches of an MRI system are shown in Fig. 7.8.

The biggest problem associated with MRI full body scanning is the loudness of noise which the patient inside the tunnel experiences. This noise is produced by the change in the physical size of gradients and other coils when high currents are passed/ stopped through them (large currents produce heating that increases the size of coils).

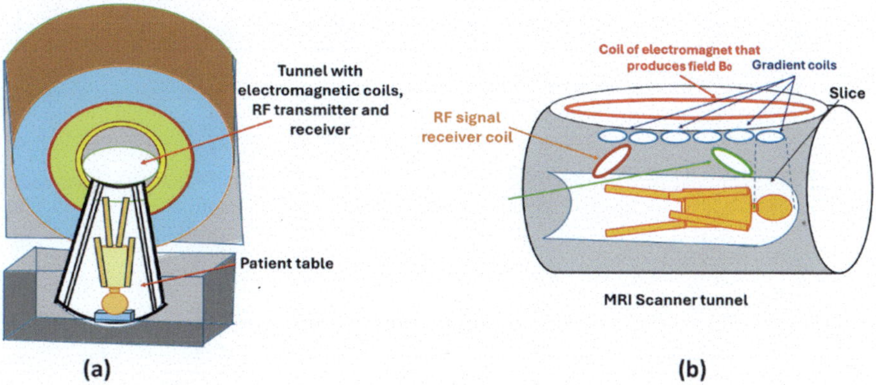

Fig. 7.8 **a** Rough sketch of the front view of MRI full body scanner, **b** rough sketch of the transfers view of the MRI tunnel

Patients are given earplugs to reduce the discomfort from background noise. Often patients feel discomfort while they are put in the tunnel, they are provided a remote button to indicate the system operator if they are not feeling well.

Dyes to enhance color contrast in the image are often used in MRI technique also. The biggest advantage of the MRI technique (over CAT and CT scan) is that it is noninvasive and that living cells are not adversely affected in this method as the RF waves used in this method are non-ionizing. The big disadvantage of this method is that it is not capable of taking images of bones and teeth that have fewer protons and higher relaxation times that may run in minutes.

Chapter 8
Positron Emission Tomography (PET Scan)

Abstract Positron emission tomography (Pet scan) is a beautiful example as to how a typical nuclear process, 'positron annihilation', may be used to identify human body parts of higher metabolic activity. It has been observed that the metabolic activity (sugar to energy conversion pathways) of abnormal body parts (like the sites of cancer, brain parts with abnormal neuroactivity, etc.) consumes sugar at a faster rate than the corresponding normal part. As a result if sugar is injected in the body, it accumulates at these abnormal activity locations in the body. Sugar loaded with some short-lived radioactive (tracer) material that emits positrons in decay is injected in the body through an intravenous line. The externally injected tracer sugar accumulates at sites of abnormal metabolic activities in about half an hour. The body is then scanned for the pair of annihilation gamma rays of 511 keV energy emitted in opposite direction in the process of electron—positron annihilation at the site of the abnormal activity. All details of the method are given in this chapter.

8.1 Introduction

Positron emission tomography is the technique to identify and image an anatomic body part by studying the performance level of its ***metabolic*** and/or ***biological*** processes. Humans eat food, digest it and convert it primarily into glucose, amino acids (that make up proteins) and fatty acids (that make up fats). The digested food: glucose, etc., is absorbed in the blood and is distributed to different parts of the body through blood circulation. Tissues and cells in different body parts (anatomic bodies) convert glucose/sugars into energy; energy is defined as the ability to do work. It may not be wrong to say that modern civilization has become possible because humans learned how to change energy from one form to another and then to use it to do work. Energy is something that a physical system may possess, and work is something that an energetic system may perform. As such both the energy and the work are measured in the same units called ***Joule*** represented by the letter **J**. There are other units also, like erg, electron volt (eV), etc., which are used to measure energy/work.

R. Prasad, *Understanding Medical Devices*, Series in BioEngineering,
https://doi.org/10.1007/978-3-031-88630-0_8

The pathways in the human body that convert food to energy constitute **metabolic *activity***. Metabolism refers to the chemical (metabolic) processes that take place as the human body converts foods and drinks into energy. Metabolism in the human body constantly provides it with energy for essential body functions like breathing, circulating blood, growing and repairing cells, managing hormone levels, regulating body temperature and digesting food, etc. The human body needs a minimum number of joules (of energy) to sustain these essential functions; this minimum number of joules is called the ***Basal Metabolic Rate*** (***BMR***). BMR value varies from person to person. An average man has a BMR of around 7100 kilojoule (kJ) per day, while the average woman has a BMR of around 5900 kJ per day.

It has been observed that the metabolic activity of abnormal body parts, for example the anatomic part that has cancer, ***increases much more than*** that of the corresponding normal part. Normal cells have a particular way of converting carbohydrates, fats and proteins into energy; however, cancer cells do it differently. Compared to healthy cells, cancer cells use more glucose to produce the same amount of energy; therefore, consumption rate of glucose in cancer infected body parts is much higher than the corresponding healthy parts. Similarly, the metabolic activity of parts of the brain that do not function properly get considerably enhanced as compared to the part that has a normal function. As a result of the enhanced metabolic activity, a considerably larger amount of sugar/glucose gets accumulated in the abnormal body part. In PET scanning, the patient is given some form of sugar like glucose that has a very small amount of some short-lived radioactive atoms attached to it. Such sugar is called ***tracer labeled sugar*** and the short-lived radioactive material as ***tracer***. The glucose labeled with radioactive tracer atoms is generally injected to the patient in a vein by an intravenous (IV) line. In a time from 30 to 90 min, the injected tracer-labeled glucose gets distributed in the patient's body and (glucose with radioactive atoms) accumulates in the abnormal body part in larger concentration where the radioactive atoms undergo decay. The decaying radioactive atoms of tracer emit radiation in the abnormal body part where they have accumulated in large quantities. Radiation emitted from the abnormal body part is recorded by suitable detectors held in the scanner of the system. Detectors convey the information obtained by them to the computer which analyzes the data and illuminates the (abnormal) area of the body where the radioactive material has accumulated, producing the required image of the abnormal body part. The intensity of the image depends on the decay rate of tracer atoms; larger the number of tracer atoms accumulated in the affected part, more will be the metabolic activity of the part corresponding to a higher level of abnormality.

8.2 Radioactivity and Tracers

Some atoms in nature are found to be unstable and undergo spontaneous disintegration. This property of disintegration is called ***natural radioactivity***. Essentially it is the nucleus of a radioactive atom that disintegrates; hence radioactivity is a nuclear phenomenon. If one takes some atoms of a radioactive material, the number of radioactive atoms decreases with time; the time in which the number of radioactive atoms reduces to half of its original number is called the ***half-life*** of the atom. Radioactive atoms with half-lives of several thousand and billions of years to a few milli-microseconds are found in nature. It is also possible to create new radioactive atoms by hitting a sample of stable atoms by focused beams of high-energy nuclear particles like neutrons, protons, alpha particles, etc. Nuclear projectiles may be accelerated to high energies in special machines called particle accelerators or atom smashers.

An important property of radioactive atoms is that each atom of the given species decays by a very specific decay mode and that the decay modes of atoms of two different species are different from each other. The decay mode of a radioactive atom is like the fingerprint of that atom. This property of individual decay modes makes it possible to identify the radioactive atom in a unique way, without any ambiguity, by studying its decay mode. Radioactive atoms when decay emit specific radiation, like alpha particles, beta particles and/or gamma rays. The alpha particle is the nucleus of the Helium atom, while there may be two types of beta particles, the beta-plus denoted as β^+ and beta-minus, denoted by β^-. Gamma rays, high-energy electromagnetic radiations denoted by γ, are also emitted in radioactive decay.

Radioactive atoms of short half-lives are often used for research on living beings and plants. A small quantity of these short half-life radioactive atoms, called ***tracers***, or ***radiopharmaceuticals***, is mixed with a suitable substance, a fluid, and is injected into the living object. The flow of the injected substance/fluid in the object may be monitored by detecting the specific radiation emitted by the decaying radioactive tracer atoms. The reason why atoms of short half-life are used as tracers is the fact all the radioactive tracer atoms of short half-life injected in the living object die out in a short time. It is estimated that in a time span equal to about five times the half-life of the atom, almost 99.9% of injected atoms die out.

In PET scanning radioactive atoms of ^{18}F (an isotope of element Fluorine) are mixed with glucose as tracer atoms. The ^{18}F tracer mixed glucose is called ***Fluorodeoxyglucose*** or **FDG**. The half-life of ^{18}F is about 110 min, and therefore, almost all atoms of ^{18}F die out of the body within say 10 h of their injection to the body. When an atom of ^{18}F decays, it emit a nuclear particle called ***positron*** that is denoted by the symbol β^+. It is like an ***electron*** that revolves around the nucleus in atoms, with the **difference** that while the electron (denoted by $\mathbf{e}^-$ or β^-) has a negative electric charge, the positron has an equal amount of positive electric charge. Positron is also called the antiparticle of an electron. In particle physics, there are several pairs of particles and their antiparticles.

8.3 Annihilation Gamma Rays

Since the metabolic activity of the abnormal body part (infected with cancer or not working normally) is higher than the normal part, a considerably large amount of ^{18}F activated sugar injected in the blood through IV line gets accumulated in the cancerous or abnormal body part. The tracer atoms of ^{18}F, accumulated in the abnormal part, undergo radioactive decay emitting ***one*** positron (β^{+}) per decay in the abnormal body part. The emitted positron moves a very small distance, less than 1 mm, when it encounters an electron (β^{-}) of some atom present in this part. Positrons (emitted by tracer atoms) and electrons (present in atoms in the body part) undergo a very special phenomenon called '***Annihilation***' when they meet each other. In the process of annihilation, the positron and the electron vanish (get converted in energy) and two gamma radiations each of energy 511 keV are emitted ***simultaneously in opposite directions***. The angle between the emitted gamma pair will be 180° if they annihilate at rest (i.e., the process of annihilation takes place when both the positron and the electron are at rest or stationary; however, in most cases, the annihilation occurs while they are still in motion) but will be slightly less than 180° (may be 178°) if the annihilation takes place when they (the positron and electron) are still moving. Figure 8.1 shows the decay sequence of the ^{18}F atom and annihilation of the electron positron pair. It may be noted that the sizes of atoms, the positron and the electron are highly exaggerated in the figure. A few pairs of gamma ray detectors, placed opposite to each other, are kept in the scanner of the PET system. The gamma detectors of each pair are tuned to detect annihilation gammas of 511 keV energy and that reach the two detectors within a very short time of the order of 10^{-9} s. The selection of gamma ray energy and time coincidences between the two gamma rays detected by the two detectors, which ensures that the possibility of detecting stray gamma rays which are present in the background and are not associated with annihilation of positron–electron pair is reduced to almost zero; the gammas detected by the two detectors are actually the annihilation gammas created as a result of annihilation of the positron–electron pair. The detectors generate signals only for 511 keV gamma rays and that are in time coincidence in the two detectors placed diagonally opposite to each other. Detectors provide coincident gamma signals to the computer of the system. The computer analyzes the detector data and by extending the trajectories of detected gammas in the backward direction, with the help of suitable software, determines the point on the patient's body where the event of annihilation has taken place. Further, the computer illuminates these annihilation points. Since there is excess accumulation of ^{18}F loaded sugar in the abnormal/infected body part, many positron–electron annihilation events take place in that part that are detected by coincidence gamma detectors and the part is brightly illuminated by the computer in comparison to the other normally functioning body parts (where only few annihilation events occur). It may be noted that in actual PET images, the abnormal/infected body parts appear darker, since they are negative of the actual photograph. In short, PET scanning is the technique of mapping the 3-D tracer concentration in the body using the positron–electron annihilation process.

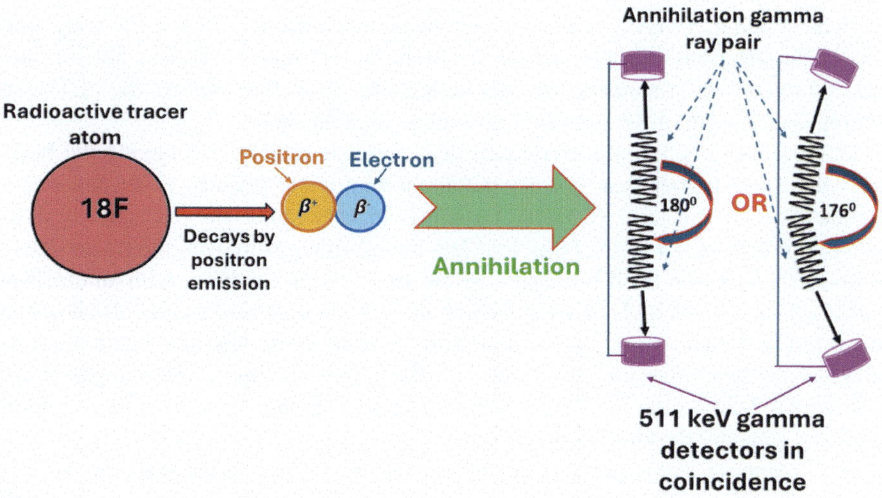

Fig. 8.1 Pictorial representation of ^{18}F decay and gamma annihilation

8.4 PET Scanner Unit

A PET scan machine, shown in Fig. 8.2, has four important components: (a) a control console, (b) a powerful minicomputer with appropriate analyzing software, (c) a doughnut-shaped scanner unit that has a large hole at its center and (d) a patient bed that may slide into the central hole of the scanner unit. The scanner unit may rotate at different speeds around the patient's bed. There are a few (up to four) detector setups fixed in the scanner unit spaced symmetrically. These detector setups rotate around the patent's bed with the rotation of the scanner unit. Each detector setup consists of a pair of gamma ray detectors held diametrically opposite to each other to detect the pair of coincident annihilation gamma rays emitted when a positron (emitted by the tracer atom) combines with the electron (present in human body). Events (simultaneous detection of 511 keV gammas by the two detectors of a given detector setup) recorded by different detector setups activate the analyzing software of the minicomputer of the system that extrapolate the trajectories of the detected gamma rays to identify and illuminate the point on human body where the annihilation event has taken place. The computer also takes the snapshot of the illuminated area. In each rotation of the scanner, several snapshots of the part of the patient's body ***lying immediately below the rotating scanner*** with illuminated area indicating the locations with large FDG concentrations are taken by different detector setups. The analyzing software generates a single image synthesizing the large number of snapshots taken by different detector setups in one rotation. This image is called a ***slice***. With the patient lying on the bed, the operator/technician operating the system remotely from the control console may bring different sections of the patient's body under the scanner one after the other and may generate several slices of the target body part. A full-body image is made by synthesizing images/slices of different sections. A full-body PET scan is of great help in identifying regions of enhanced metabolic activities in the body, that may be the locations of high potential for future development of cancer, etc., or may indicate the on-set of abnormal metabolic activity. To further confirm the PET scan findings, a CT scan of the patient is also taken in most cases.

PET scanning is a noninvasive, harmless and safe method of imaging the body parts that have higher than normal metabolic/biological activity. Since the tracer radioactive material injected in the patient is in very small amounts and of short half-life, it decays out of the body in a few hours' time. Moreover, the tracer decay product; the positron itself has a small-time span, and it combines with an electron within a few nanoseconds of its emission, as such there is no residual effect of the tracer left in the patient in a few hours' time. During these few hours also, there is no adverse effect either on the patient or to the people staying near to the patient.

Figure 8.3 shows the FDG PET scan of a patient having a cancer-infected lymph node and a cancer developing in the neck.

PET scanning technique is a beautiful example of how a nuclear process, the annihilation of a positron and an electron, has been used to locate body parts with abnormal metabolic activity.

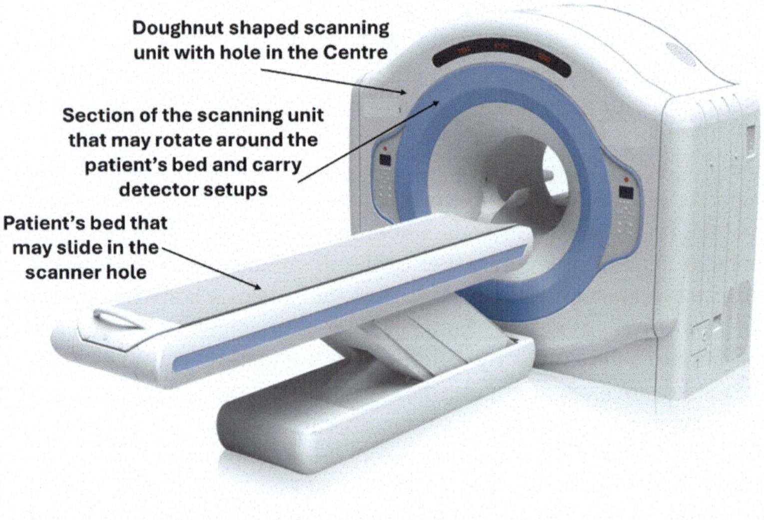

Fig. 8.2 Image of a PET scan unit

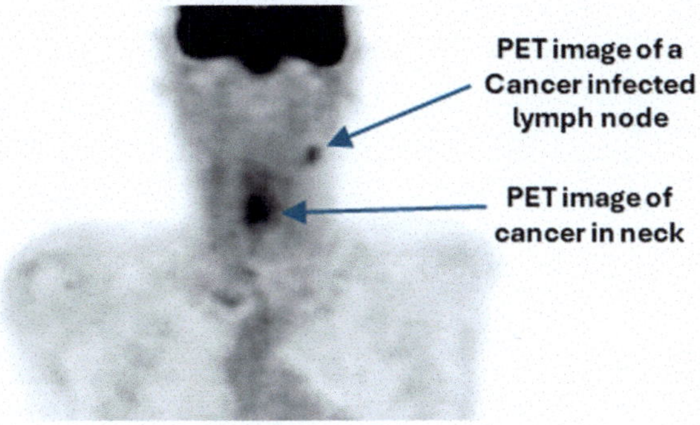

Fig. 8.3 FDG PET scan of neck indicating the infected lymph node and a part of neck

Chapter 9
Ultrasound Scan

Abstract Sound waves with frequencies above 20 kHz are called ultrasound. Ultrasound is frequently used to decipher the underground and underwater structures. Ultrasound signals of appropriate frequency are now often used to generate images of internal body organs, in determining their size and the depth from the body skin. Since ultrasound signals are non-toxic and non-ionizing, they do not damage or harm body cells; one may therefore take several ultrasound scans without any damage to the body. Using the Doppler effect, ultrasound signals may also be used to generate the image of blood flow pattern in different body parts. Details of ultrasound imaging are provided in this chapter.

9.1 Introduction

Sound waves are produced by the vibrations of an object in an elastic medium. For example, when the arm of a tuning fork (an instrument used to produce sounds in laboratories) is struck by a hammer, it starts vibrating back and forth (or to and fro) in air as shown on the left side of Fig. 9.1. The vibrations of the tuning fork's arm displace the medium particles (air in this case) that are in immediate contact with the arm. Displaced medium particles (air molecules in the present case) try to come back to their original positions as they are held together by elastic forces. Elastic force is that force which brings a given material back to its original shape after being deformed. Back and forth motion of the tuning fork's arm deforms or disturbs the medium (air) particles or molecules, but the elastic force between air molecules (elastic medium) forces the displaced air molecules back towards their original positions. Mediums, the constituents (molecules) of which are bound by ***elastic forces*** are called ***elastic media***. Many gases, most of the fluids including water and many solids, metals in particular, are all elastic media. The characteristic property of the elastic force is that if a particle held by elastic forces is displaced from its initial position (called the equilibrium position) within some limits, the elastic force pulls the particle back towards the equilibrium position. However, the particle while coming back develops some speed, because of the elastic force of pull, and overshoots the equilibrium

R. Prasad, *Understanding Medical Devices*, Series in BioEngineering,
https://doi.org/10.1007/978-3-031-88630-0_9

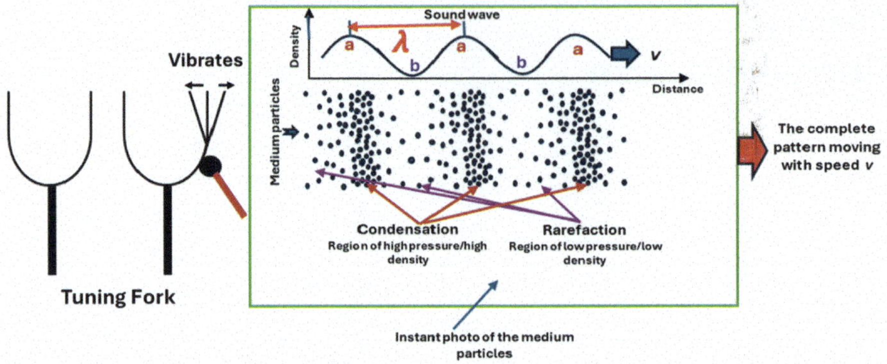

Fig. 9.1 When an arm of a tuning fork is hit by a hammer, it starts vibrating, producing sound waves in the surrounding medium (air in this case). Sound waves consist of alternate regions of condensations and rarefactions

position onto the other side. In this way, the particles of an elastic medium, when displaced, undergo vibratory motion about their equilibrium position.

Vibrations of the arm of the tuning fork produce displacement of the medium particles setting in a sound wave in the elastic medium consisting of alternate regions of higher and lower concentrations of medium particles as shown in Fig. 9.1. The region where the concentration or density (number of medium particles per unit volume) is higher is called the **condensation** and the one with lower concentration the ***rarefaction***. Also, the pressure is higher in the region of condensation and lower in the region of rarefaction. A sound wave in an elastic medium may, therefore, be visualized as alternate regions of higher and lower particle densities or pressures. The right-hand side of Fig. 9.1 shows a snapshot (photo at a given instant, say '*t*') of the sound wave at some instant '*t*'. If one plots the particle density as a function of distance in the wave, one gets the curve shown at the top of Fig. 9.1. A sound wave is often represented by this curve in figures. The points marked by 'a' in the figure are called ***crest*** and those by 'b' as ***trough***; particle density or pressure in the wave is highest at crests and minimum at troughs. The distance between two successive crests or troughs is defined as the ***wavelength*** of the sound wave and is denoted by the Greek letter λ. It is obvious that if there is no medium and specially, no elastic medium around the vibrating tuning fork, no sound waves will be produced and propagated; therefore, ***an elastic medium is essential for the production and propagation of sound waves***.

The sound waves produced by a vibrating object (tuning fork in the present case) move away from the source of sound through the elastic medium (air in this case) with a speed V. The propagation of the sound wave means that the whole pattern consisting of crests and troughs (or condensation and rarefactions) moves away from the source, the tuning fork, with some speed V. It essentially means that the positions of crests and troughs in a sound wave, which is moving in a medium, do not remain fixed. The complete pattern of crests and troughs moves through the medium with the velocity $\boldsymbol{V}$ of the wave as shown in Fig. 9.1. It means that if at a particular point in the medium there was a trough of the moving sound wave at some instant 't', then at some later instant 't_1' there may be a crest at the same point. In a moving sound wave, crests and troughs keep passing through a particular point in the path of the wave. The distance traveled by a given trough or crest in unit time is called the speed of the sound waves in that medium. The sound waves travel with the speed of 343 m/s in air at 20 °C temperature, with 1481 m/s in water at the same temperature (almost 4.3 times as fast) and at 5120 m/s in iron (almost 15 times that of in air). Speed of sound in a medium strongly depends on the elastic properties of the medium and the temperature of the medium.

There are three important parameters associated with a sound wave: the ***frequency*** $\boldsymbol{\nu}$, the wavelength $\boldsymbol{\lambda}$ and the speed $\boldsymbol{V}$. Here frequency $\boldsymbol{\nu}$ is the property of the source, the tuning fork, and gives the number of vibrations that the vibrating object (tuning fork arm) executes in unit time of one second. If the source of sound is not changed, the frequency $\boldsymbol{\nu}$ remains constant; it does not depend on the properties of the elastic medium or the temperature. The velocity of the sound wave $\boldsymbol{V}$ and the wavelength $\boldsymbol{\lambda}$ booth depends on the medium and its temperature. However, $\boldsymbol{\nu}$, $\boldsymbol{\lambda}$ and $\boldsymbol{V}$ at a given temperature T are related to each other by the following relation:

$$V = \nu \cdot \lambda \text{ (at a fixed temperature).} \tag{9.1}$$

Equation 9.1 tells that for a given medium and at a fixed temperature, the wavelength of the sound wave is inversely proportional to the frequency of the source.

Since the frequency of the sound wave depends only on the source and is independent of the medium and the temperature, sounds are classified in terms of their source frequency. Further, the frequency of the sound is also called the ***pitch*** of the sound. Sounds produced by sources that vibrate with frequencies between 20 vibrations per second (20 Hz, twenty Hertz) and 20,000 vibrations per second (20 kHz, twenty kilohertz) are called ***Audio sounds*** as the human ear can detect them. Sounds of frequencies less than 20 Hz are called ***subsonic***, which the human ear cannot detect but some animals, like elephants, may hear them. Sounds of frequencies larger than 20 kHz called ***Ultrasonics, supersonics or ultrasound***, though not audible to humans, may be detected by some birds and reptiles. It is believed that bats and snakes generate ultrasound signals and navigate their path by sending ultrasound signals and picking back the signals called ***echo***, reflected from obstacles in the path.

Essentially, waves are means of transferring energy from one place to another. For example, electromagnetic waves (Sunlight) coming from the sun bring heat energy with them. Similarly, sound waves carry energy with them as they travel through a medium. The energy content of a sound wave depends on two factors (i) the ***amplitude***; the maximum displacement of the vibrating body from its equilibrium position and (ii) the frequency of the sound wave. The intensity (denoted by letter I) of a sound wave is defined as the energy that passes through a unit area, held perpendicular to the direction of the propagation of sound wave, per second. In MKS system, the unit of area is meter^2 (denoted as m^2) and that of energy per second is Watt (denoted as W); hence,

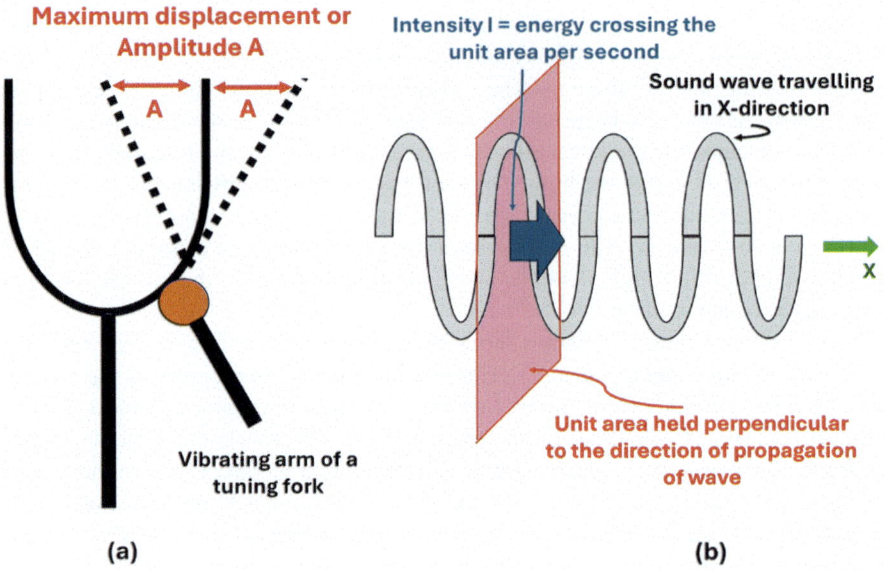

Fig. 9.2 **a** Maximum displacement of a vibrating body from its mean position is called the amplitude, denoted by A, **b** intensity of a sound wave is defined as the energy crossing a unit area held perpendicular to the direction of propagation of the wave, per second

$$I = W/m^2 \text{ Watt per square meter.} \tag{9.2}$$

Concepts of the amplitude and the intensity of a sound wave are shown in Fig. 9.2.

Another important parameter associated with the intensity of sound is ***Loudness***; it relates the intensity of any sound to the intensity at the threshold of hearing. Loudness is measured in units of ***decibel*** denoted by db. The loudness of sound depends both on the intensity and the distance of the observer from the source. The threshold of human hearing has an intensity of 1×10^{-14} watts per meter square. The loudness of the sound of normal breathing is about 10 db; of normal conversation is about 60 db and of noise more than 80 db.

Ultrasound waves carry higher energy density as compared to audio waves. Energy of sound waves dissipates as it travels through some medium. However, ultrasound may travel considerably longer distances in each medium as compared to the audio or subsonic waves. It is for this reason that ultrasound signals are used in determining the depth of sea or of sunken ships and other objects, etc. This branch of ultrasonic application is called SONAR, an abbreviation for Sound Navigation and Ranging. Very powerful ultrasound signals are now being used to decipher the hidden structures buried deep under the earth surface.

Like the light rays (electromagnetic waves), sound waves traveling in a medium get reflected from the edges and boundaries of objects or obstacles they come across in the medium. Reflection of sound waves follow the well-known laws of reflection; the angle of reflection is equal to the angle of incidence and the incident ray, the normal to the reflecting surface at the point of incidence and the reflected ray lie in the same plane. This is shown in Fig. 9.3a. Normal incidence of ultrasound waves is often used to determine the distance of an object/obstacle from the source of the ultrasound. In this procedure a short burst of a focused ultrasound signal is directed towards the object. The ultrasound pulse gets reflected from the object and traveling back along the path of the incident, which reaches an ultrasound detector. The time '*T*' taken by the ultrasound pulse in going and coming back after reflection from the object is recorded. If *V* is the speed of ultrasound in the medium (*V* is known), and *D* the distance of the object from the detector, since distance traveled (2D) = speed (*V*) × time (*T*); therefore, $D = V \cdot T/2$. Thus, with the known value of the speed *V* of ultrasound in the given medium and the measured value of time *T* that elapses between the sending of ultrasound signal and receiving the reflected signal, it is possible to determine the distance of the object from the detector/source of ultrasound.

When a pulse of some ultrasound wave hits the boundary of an obstacle, a part of the wave is reflected back, and the intensity of the reflected ultrasound signal carries information about the density of the material of the obstacle. The intensity of the reflected signal is higher if the density of the obstacle material is high. This property of the reflected ultrasound signal is used to decipher the nature and type of the reflecting object. The intensity of the echo signal is higher if the reflecting object has high density and vice versa.

Doppler effect It is a common experience that the pitch (or frequency) of the whistle blown out by a moving train appears to change as the train approaches a stationary observer standing on the railway platform; the pitch appears to increase as the train approaches and the pitch appears to decrease as the train passed the observer and

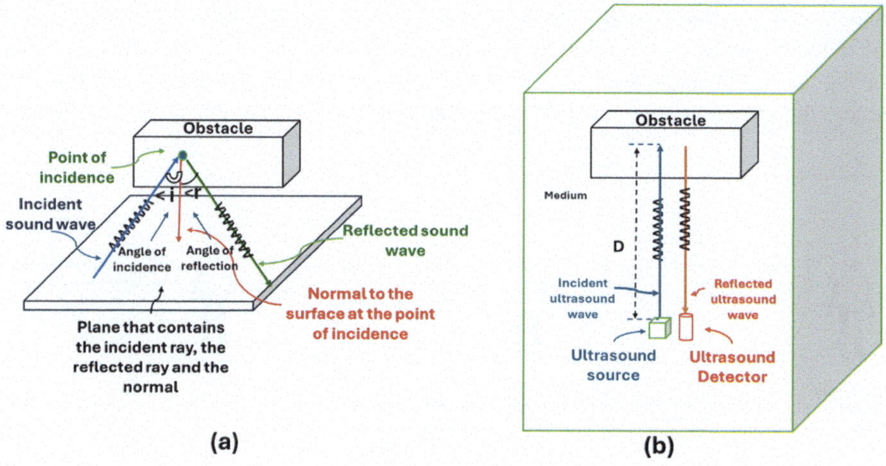

Fig. 9.3 **a** Pictorial representation of reflection of ultrasound wave by an obstacle surface, **b** determining the distance of the obstacle using the normal incidence of ultrasound pulse

recedes away from him. This change in the frequency of the whistle is only apparent, in fact the frequency of the train whistle is fixed, but because of the relative motion of the source (train) and the observer (man), the sound frequency appears different from the actual frequency. This apparent change in the frequency of a wave due to relative motion of the source and observer is called the ***Doppler effect***. The Doppler effect was first described by the Austrian physicist Christian Doppler in 1842. It takes place both for mechanical (sound) and electromagnetic (light) waves.

Without going into the physics of the effect, the basis of the process may be explained with reference to Fig. 9.4, where, on the left side of the figure, a stationary vehicle produces a sound signal of frequency f_0. The signal moves out in the form of sound waves in the air in all directions. Two stationary observers, A and B in opposite directions, hear the sound signals and both detect the frequency of the signal as f_0 since the signal is traveling with the same speed in air in both directions. On the right side in Fig. 9.4, it is shown that the sound of the same frequency f_0 is produced by the vehicle which is moving towards a stationary observer marked as C. As may be seen in the figure, the sound waves emitted from the moving vehicle compresses themselves, apparently reducing the wavelength of sound waves and, therefore, observer C finds the wavelength of sound reaching him shortened and the frequency increased. However, observer B from which the source of sound (vehicle) is receding observes a sound wave of longer wavelength (and lower frequency) approaching him.

In Fig. 9.5 a, b, c it is shown that an ultrasound signal of frequency, say, f_0 is projected in a medium and hits an obstacle along the normal to its surface. The signal gets reflected from the surface and moves in the medium towards a stationary detector. Since the reflected signal is coming from the surface of the obstacle, it is reasonable to assume that the obstacle surface is the source of the reflected ultrasound signal of frequency f_0. In Fig. 9.5a, both the source (the obstacle surface) and the detector are stationary and, therefore, there is no Doppler shift in frequency, and the detector observes the frequency of the signal as f_0. In Fig. 9.5b, the source of the signal, the obstacle surface, is moving with speed V_1 towards the detector, and hence, the apparent frequency of the signal observed by the detector is f_1, that is

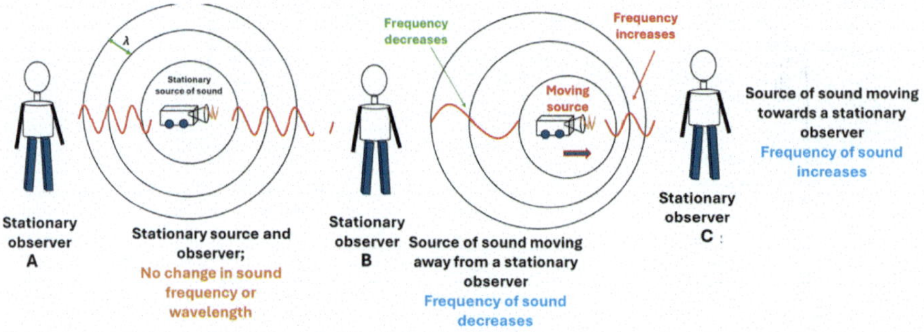

Fig. 9.4 Pictorial representation of Doppler effect

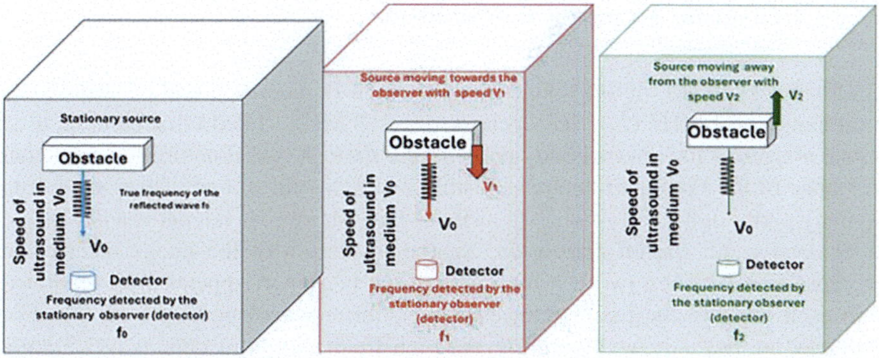

Fig. 9.5 **a** Signal reflected from the surface of the stationary obstacle moves with the speed V_0 in the medium and is observed by the stationary detector as of original frequency of f_0. **b** Signal reflected from the surface of the obstacle that is moving towards the stationary detector with speed V_1 travels in the medium with speed V_0 and appears to have a higher frequency f_1. **c** Signal reflected from the surface of the obstacle which is moving away from the detector with the speed V_2 travels through the medium with the speed V_0 and is detected by the stationary detector to have the lower frequency f_2

higher than f_0 $(f_1 > f_0)$. In Fig. 9.5c, the source of ultrasound signal is receding from the stationary detector with the speed V_2 and, therefore, the detector registers the ultrasound signal of apparent frequency f_2 (instead of f_0) because of the Doppler effect $(f_2 < f_0)$. In all the three cases the signal moves in the medium between the obstacle and detector with the speed V_0. Expressions for the apparent frequencies f_1 and f_2 may be written as:

$$f_1 = [v_0/(v_0 - v_1)]f_0 \tag{9.3}$$

and

$$f_2 = [v_0/(v_0 + v_2)]f_0. \tag{9.4}$$

As may be seen in Eq. 9.3 the quantity in the bracket is larger than 1, and hence $f_1 > f_0$; while in Eq. 9.4 the quantity in the bracket is smaller than 1, therefore, $f_2 < f_0$. The quantities $(f_1 - f_0)$ or $(f_0 - f_2)$ are called ***Doppler shifts***. It may be remembered that the above expressions hold only when the reflected wave is traveling in the direction of the normal to the obstacle surface and the detector is stationary. In a more involved cases of oblique propagation of ultrasound wave and both the source and the detector in motion, the expressions for apparent frequencies are more complicated and contain other terms.

Equations 9.3 and 9.4 also tell that if f_0, f_1 (or f_2) and V_0 are known, it is possible to calculate the speed V_1 or V_2, the speed with which the reflecting obstacle is moving and in which direction with respect to the detector.

9.2 Ultrasound Imaging Kit/Scanner

As already explained, in ultrasound imaging, high-frequency sound of frequencies in the range of 2 MHz (2×10^6 cycles/sec) to 18 MHz, called ultrasound, is used to take images of the internal body parts in ***real time***. A real-time image means that the image of the body part becomes visible while the ultrasound waves are hitting the body part. Higher ultrasound frequencies have shorter wavelengths and provide better details and ***spatial resolution***. Spatial resolution of the image means that the separation between two nearby points of the body part appears well separated from each other in the image. However, the shorter wavelengths are absorbed or attenuated more easily, and therefore, the high-frequency ultrasound waves are not so penetrating. Medical ultrasound machines, therefore, use ultrasounds of different frequencies for imaging different body parts. For example, the ultrasound of highest frequencies (18–15 MHz), which are least penetrating, is used for imaging superficial structures, musculoskeletal, etc. 10–7 MHz ultrasound is used for taking images of breast, thyroid, superficial veins etc. Ultrasound signals in the range of 5–2 MHz are employed in imaging of pelvic region, general and deep abdomen, gynecological organs, etc.

Two big advantages of ultrasound imaging are: (i) Apart from being non-invasive, the ultrasound waves used in this imaging are non-ionizing; hence they do not produce any toxic or radiation side effects. It is even safe to use ultrasound imaging to monitor the growth of the fetus in the mother's womb. (ii) Since ultrasound imaging is done in real time, several images of the same body part may be taken successively after short intervals to detect time-dependent changes in the structure of the target body part. Not only that, blood flow in veins, arteries, or heart chambers may be monitored using the Doppler effect for ultrasound signals reflected from moving red blood cells.

9.3 Ultrasound Scanning System

There are two important components of a diagnostic ultrasound scanning system: (i) a ***handheld transducer or probe*** and (ii) ***a powerful computer***. The transducer is generally a small plastic box-like structure with a narrow opening for sending and receiving ultrasound signals and houses (a) either a single source or an array of sources of ultrasound signals and (b) either single or an array of ultrasound detectors. Cables to provide electric power to ultrasound source(s) and detector(s) and to transfer information acquired by detectors, etc., to the computer, are all taken from the back of the transducer to the computer and to electric power supplies located in a small box placed nearby.

Ultrasound Source Ultrasound signals are short duration packets of high-intensity ultrasound waves of known frequency that are focused on a particular direction. The distance of the point of focus of the signal from the source can be controlled by adjusting the parameters of the focusing unit. Ultrasound sources are generally made of a special type of piezoelectric–quartz crystals. Piezoelectric crystals have a unique property that when an electric potential pulse is applied across a given pair of opposite faces of the crystal, the other pair of opposite faces of the crystal starts vibrating with the frequency of the applied electric potential pulse. This happens because on application of an electric potential difference between two specified opposite faces of the crystal, the length of the crystal in the direction normal to those faces increases; and when the applied electric potential becomes zero, the crystal retains its original length.

Figure 9.6 shows the production and focusing systems for the generation of ultrasound signals. An electronic unit called multivibrator produces a train of electric pulses of voltage x and time duration T as shown in the top part of Fig. 9.6. This electronic pulse train is applied to the specific faces A and B of the piezoelectric crystal. The length of the crystal increases in the direction of faces C and D when potential X is applied between faces A and B; however, the crystal length reverts to the original value when potential X becomes zero. Thus, with the application of the voltage pulse train between faces A and B of the crystal, the other faces C and D of the crystal are set in vibratory motion. When the frequency of the electric pulse train is in the range of ultrasound frequency, ultrasonic waves are generated. Ultrasound waves are, therefore, produced in the surrounding air medium. The frequency of the ultrasound signal may be controlled by controlling the repetition rate of positive voltage pulses produced by multivibrator, while the intensity of the ultrasound signal may be controlled by the amplitude X of the positive pulse. The ultrasound signal may be focused to the desired depth using a suitable focusing device called lens. Focusing lens may be an electronic unit. The ***handheld transducer unit*** (see Fig. 9.7) of the ultrasound scanner also contains detectors that may detect and record the intensity of the ultrasound echo reflected from the boundary of some internal structure of the body. In principle a tiny microphone may work as the ultrasound echo detector; it converts the ultrasound pressure wave into an electric pulse, and the amplitude of the electric pulse is being proportional to the intensity of the received echo signal. Two important characteristics of the detected echo pulse are recorded by the detector; the time it took to reach back and the intensity of the pulse. This information about the characteristics of each recorded echo pulse is sent to the computer that analyzes them and creates the image of the structure the boundary of which reflects the recorded echo pulses.

9.4 Scanning Modes

An ultrasound scanning unit may be operated in four different modes.

Mode-A: It is the simplest mode of ultrasound operation. In this mode, only a single ultrasound source in the transducer sends signals along a line in the body and plots the intensity of echo as a function of depth, to study the structure of a specific tumor or calculus.

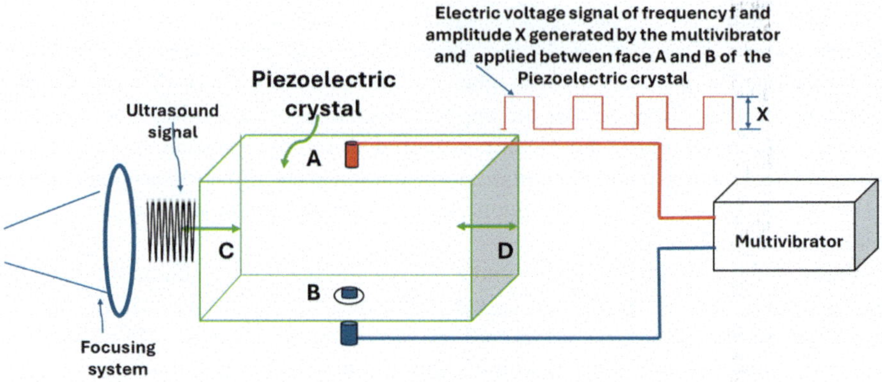

Fig. 9.6 Schematic diagram of generation and focusing of ultrasound signal

Fig. 9.7 Handheld transducer/probe unit

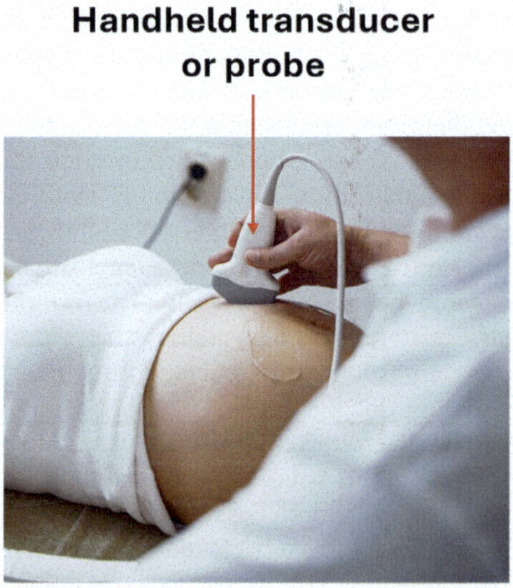

Mode-B: In this mode, a linear array of sources and detectors in the transducer simultaneously scan a plane through the body and produce a two-dimensional image of the body part on the screen.

Mode-M: The letter M in this mode stands for motion. The two-dimensional scan of a body part taken in Mode-B is repeated after a short time interval several times and displayed on the screen to study the time-dependent changes occurring in the part.

Doppler Mode: In this mode Doppler shifts in the frequency of the incident ultrasound signal and the reflected signal from the moving blood cell, is determined to visualize the blood flow. With the knowledge of the frequency of the incident ultrasound signals, speed of sound in the medium (body part) and the frequency of the ultrasound signals reflected from the moving blood cells (and detected by detectors in the handheld transducer), the computer of the scanner may calculate the speed and the direction of flow of blood cells. By calculating the Doppler frequency shift of a particular sample volume, for example a jet of blood flow over a heart valve, it is possible to determine or visualize the speed and the direction of flow of the blood volume. This information is particularly useful in cardiovascular studies. The computer of the scanner may display the Doppler information graphically, i.e., showing different blood volumes flowing with different speeds and in different directions by different spectral colors. This is referred to as ***spectral Doppler***. If the Doppler information is displayed as an image, the scan is called ***color Doppler***.

It is to be emphasized that like other imaging systems, the computer associated with the scanner of ultrasound kit plays the key role of generating the image. The ultrasound generator(s) and detector(s) housed in handheld transducer provide the associated computer information about the time at which the ultrasound signal was sent, the time at which the corresponding echo (or the reflected ultrasound signal) was received, the frequencies and the intensities of the incident and echo signals etc. The average speed of ultrasound in different body parts is already stored in the memory of the computer. Using this data, the computer generates the image of the body part, the boundary of which reflects the echo signal. The timing information provided by detected echo signals makes it possible to estimate the size of the internal structure. Computers may also show different structures with different shades of gray or colors, depending on the intensity of the echo signal. As such the computer of the scanner is the key to the machine that renders the visible image.

Doppler ultrasound images are often used to detect narrowed arteries and veins, blood clots and deep vein thrombosis (bleeding), blood vessel injuries, and to evaluate the blood supply to organs transplanted in the body, etc.

9.5 Preparing for an Ultrasound Scan

Though most of the ultrasounds do not require any preparation but for a few ultrasounds doctors may advise some preparations before taking the image. For example, in the case of abdominal ultrasound and abdominal ultrasound with Doppler, the patient must not eat or drink for eight hours before the imaging schedule.

In case of pelvic ultrasound of female patients, they are required to drink a sufficient amount of liquid (about 32 oz of water) about an hour before the scan to keep their bladder full.

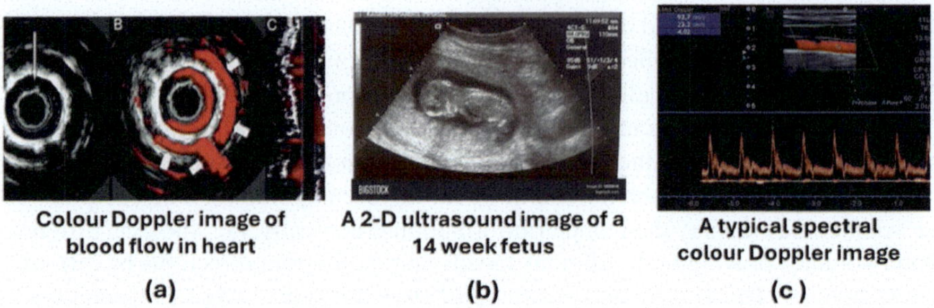

Fig. 9.8 **a** Color Doppler image of blood flow in heart, **b** ultrasound image of a 14-week-old fetus in mother's womb, **c** a typical spectral color Doppler ultrasound image

9.6 Ultrasound Procedure

In most cases, the patient is asked to lie down on an examination table or bed. The sonographer, the ultrasound technologist will put some gel on the skin of the patient above the body part that needs to be ultrasound. The gel is used to reduce reflection of ultrasound signals from the outer skin of the part thus allowing the ultrasound signals to penetrate easily through the skin into the body. The sonographer will then put the handheld transducer (also called probe) on the gel. Depending on the mode of imaging selected, the desired image of the internal structure will appear on the screen of the system, which may be a computer or a PC screen.

In some special ultrasound scanning, for example in case of the ultrasound scanning of the pelvis region of a female patient, the special transducer or probe may be inserted into the vagina (for better image) of the patient and the ultrasound image is taken.

Some ultrasound images are shown in Fig. 9.8.

Chapter 10
Glucometer

Abstract Glucometer is the device that measures the glucose (sugar) level in the blood. A substantial part of the food we eat gets converted into sugar in the process of digestion and gets mixed in the blood. A very intricate balance of the sugar level in blood is maintained by the two hormones produced by human pancreas. Breakdown of control mechanism results either in (diabetes) hyperglycemia (excess of sugar in blood) or hypoglycemia (lower sugar level), both of which are dangerous. The tight sugar control mechanism in the body, symptoms of the two types of sugar abnormalities, mechanism and method of measuring glucose blood level using different measuring instruments, etc., are discussed in this chapter.

10.1 Introduction

The digestive system of the human body converts food into three basic components: (i) glucose, (ii) fatty acids and (iii) amino acids. These components are absorbed by the walls of the intestines and are mixed in the blood. Blood carries them to the cells of different body parts and organs including muscles, liver, lungs, brain, etc. Fatty acids and amino acids are ultimately converted respectively into fat and proteins. Glucose ($C_6H_{12}O_{16}$), which is essentially a carbohydrate, has the most important function to act as a source of energy for the human body. The glucose itself does not provide energy, but it is the essential precursor in the synthesis of a compound called ***adenosine triphosphate*** (ATP) by the body cells, which is the storehouse of cell energy. Energy contained in ATP can be used to drive processes like the biosynthesis and transportation of molecules across cell membranes, etc., according to the cell requirements. Glucose can also be used in the creation of proteins, glycogen and lipids, if required by the cells. Glucose concentration in the human body fluids (blood or other fluids) is measured either in units of ***millimole per liter*** (**mmol/L**) or in ***milligram per deciliter*** (**mg/dL**). Further,

$$1\ \text{mg/dL} \approx 18 \times \text{mmol/L}. \tag{10.1}$$

R. Prasad, *Understanding Medical Devices*, Series in BioEngineering,
https://doi.org/10.1007/978-3-031-88630-0_10

10.1.1 *Natural Way of Blood Glucose Regulation in Human Body*

The blood glucose concentration in the human body is very tightly regulated, through two hormones released by the organ **pancreas**. The two hormones that have opposite effects are ***insulin*** and ***glucagon***. Insulin is produced by the beta cells while the glucagon by the alpha cells of the organ pancreas. The human body attempts to maintain homeostasis (self-regulating process) in blood glucose levels between 4 and 6 mmol/L or about 72–108 mg/dL. This glucose concentration is often called the **threshold** concentration.

In a healthy person, higher concentration of glucose in blood triggers beta cells of pancreas to generate and release insulin in the bloodstream. Insulin removes excess glucose from the blood in three different ways: (i) Insulin facilitates the absorption of glucose by muscle cells and its conversion into ATP. Energy stored in ATP may be used by cells for their various energy needs. It also works as a catalyst in the synthesis of another molecule called ***glycogen*** by muscle cells, that works as a secondary energy storage in the case of low levels of glucose in blood. (ii) In liver cells, insulin instigates the conversion of glucose into glycogen and fat. (iii) In the fat cells of the ***adipose*** tissue (a loose connective tissue composed mostly of adipocytes that provides energy storage, insulation from extreme temperatures and cushioning around soft body organs), insulin promotes the conversion of glucose into more fat. The pancreas keeps releasing insulin till the excess glucose in blood is removed by the three processes mentioned above.

When the glucose concentration in blood falls below the threshold value ($\simeq$ 72–180 mg/dL), the alpha cells of pancreas become active and release the hormone glucagon. Hormone glucagon acts on liver cells initiating the conversion of glycogen into glucose and fat into fatty acids. In this way a dynamic equilibrium of glucose concentration in blood is maintained till more glucose is fused in the blood from the fresh food intake. Figure 10.1 summarizes the processes responsible for maintaining the dynamic equilibrium of glucose level in blood.

Figure 10.1a shows how high glucose level in blood triggers pancreas to release hormone insulin and how in presence of hormone insulin the muscle cells, cells in liver and in adipose tissues convert excess glucose into glycogen and other forms of energy stores like ATP and fat. Figure 10.1b shows that below threshold glucose level in blood triggers the release of the hormone glucagon by pancreas which initiates conversion of glycogen, accumulated in liver cells, into glucose and fatty acids. The two opposite processes of removing excess glucose in presence of insulin and conversion of glycogen into glucose in presence of hormone glucagon essentially maintain the blood glucose level in the human body.

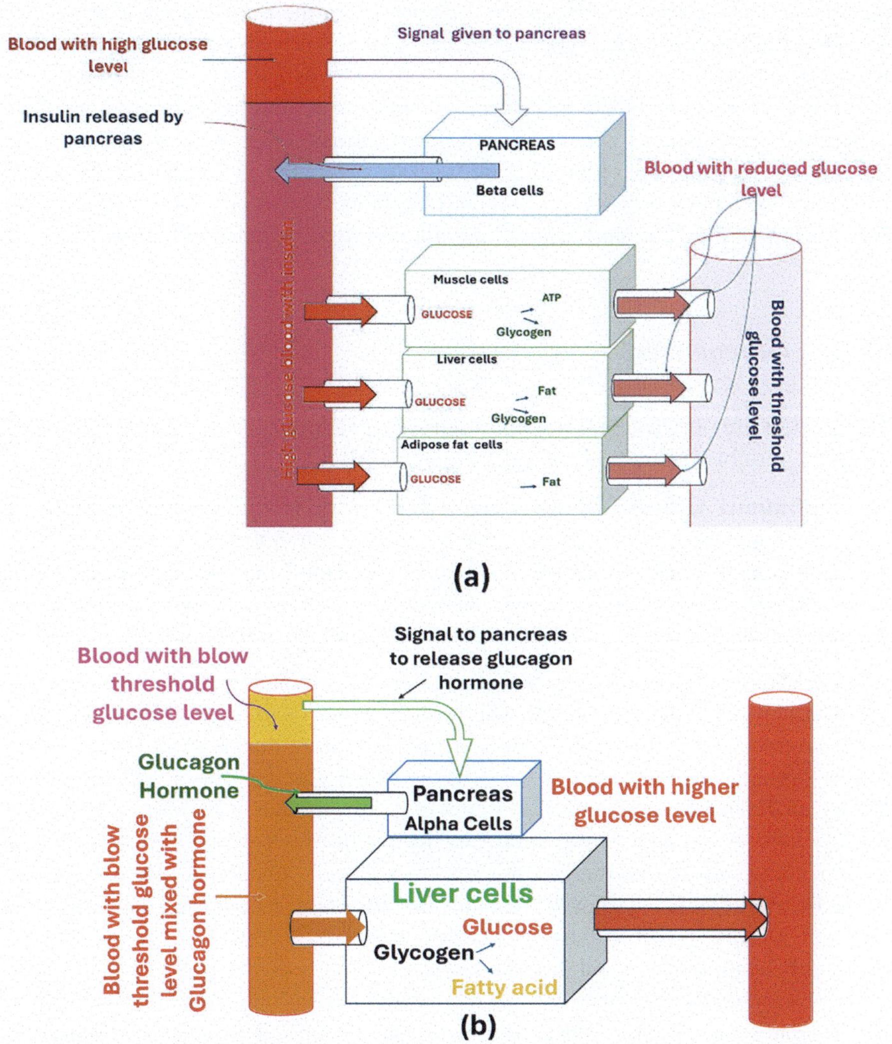

Fig. 10.1 Pictorial representation of **a** excess glucose removal from blood in presence of insulin hormone, **b** glucose enhancement in presence of glucagon hormone by liver cells

10.1.2 Uncontrolled Blood Sugar and Its Adverse Effects

Uncontrolled blood sugar is a level of glucose in blood that is outside the healthy target range. The doctor decides the healthy target range of blood glucose level for a particular patient. Blood glucose level higher than target value is called ***hyperglycemia*** and a level below target range is termed ***hypoglycemia***. The condition of hyperglycemia is generally called diabetes. People with diabetes are not able to manage their blood glucose effectively because either their pancreas does not make enough insulin (this is called type-1 diabetes) or their cells do not respond to insulin (type-2 diabetes).

Prolonged periods either of hyperglycemia or hypoglycemia may lead to dangerous consequences and may become life threatening. It is, therefore, very necessary to measure the blood glucose level occasionally for a normal person and frequently for a person diagnosed having blood sugar level out of the target range. The instrument used for measuring blood glucose level is called ***glucometer***.

10.1.3 Glucometers

Glucometers are of two types: (i) continuous glucose monitoring (CGM) and (ii) the handheld portable self-monitoring glucometer (SMG).

(i) **Continuous Glucose Monitoring (CGM)**: Continuous glucose monitoring devices are currently a major source of research on diabetes management. It is because such devices could alert a diabetes patient of impending hypoglycemic or hyperglycemic events and may also give the patient the required dose of the appropriate hormone, glucagon or insulin, as and when required.

A CGM device typically consists of three major components: (a) ***a glucose sensor*** that continuously measures the glucose level in blood or in **interstitial fluid** (fluid found around cells), (b) an ***electronic processing unit*** that is in communication (either through wires or through wireless) with the glucose sensor and (c) a ***data display unit***. The more advanced systems may also include a hormone (glucagon/insulin) delivery unit that may get activated by events of out-of-range glucose levels.

Depending on whether the glucose sensor is placed outside the patient's skin or inside, below the skin, the CGM device may be called non-invasive or totally invasive. However, it is also possible to have a partially invasive CGM when only a part of the sensor is beneath the skin and the rest stays out.

In most cases the sensor accepts a fixed volume of blood or of interstitial fluid and evaluates the glucose concentration in that volume. The glucose monitoring capability of the sensor may be based on one of the three detection technologies; (i) electrochemical, (ii) optical and (iii) other.

(i) **Electrochemical**

Glucose monitoring based on electrochemical approaches may be broadly categorized under (ia) enzymatic (using a chemical called enzyme) and (ib) non-enzymatic methods.

(ia) **Enzymatic Method**:

In this approach, the sensor is painted with an appropriate enzyme which oxidizes the glucose (contained in a fixed volume of blood or of interstitial fluid) and the enzyme itself gets reduced. The reduced enzyme may take back the oxygen either from some other chemical also painted on the sensor or from the surrounding environment, to regain its original chemical form. During the oxidation and reduction reactions, an electric current is generated between the two electrodes fixed on the sensor. The magnitude of the current is proportional to the amount of glucose in the given volume of the fluid. The current is digitized using an A/D converter and calibrated in terms of mmol/L or mg/dL or both, as required. It may be noted that after oxidizing glucose, the enzyme gets reduced, but it takes oxygen back either from the other chemical painted along with the enzyme or from the surroundings and returns to its original chemical form. Thus, the chemical state of the enzyme is restored after each analysis of blood (or fluid) samples, and it is ready for the next analysis. In short, the enzyme acts as a catalyst in the oxidation of the blood. In the process of blood oxidation, few free electrons are also produced which generate a current. The magnitude of the current is proportional to the glucose level. Most of the handheld glucometers also use the same technology, and some more details of this method are discussed under the heading of SMG.

(ib) **Non-enzymatic Method**:

In this method, enzymes are not used to initiate oxidation of blood glucose; instead, the direct electro-oxidation of glucose is achieved using **nanostructured** electrodes such as of platinum-lead alloy nanowires that exposes high surface area and work as a catalyst for the oxidation of blood. In this case, the surface of nanowires acts as the catalyst for the oxidation of blood sugar. Since wires made up of nanoparticles expose a large surface area, the process of blood oxidation becomes faster. The free electrons emitted in the oxidation of glucose coated blood constitute an electric current which is proportional to the concentration of glucose in the sample volume of the blood. In most cases the current is converted into electric voltage (voltage signal is better for further processing), which is digitized and calibrated in terms of the glucose concentration using samples of known glucose concentration in calibration experiments carried out by the manufacturing agency. Digitized voltage calibration data (voltage verses known sugar concentration) is stored in the memory of the computer for later use to determine the sugar level in an unknown sample.

(ii) **Optical Approach**:

In this method the sensor is painted with a chemical that readily binds the glucose molecules of blood and emits light of a typical color when illuminated by a very small source of invisible light (infrared or ultraviolet) contained in the sensor. The intensity of the emitted colored light is digitized and calibrated in terms of the glucose concentration. In some cases, the chemical present on the sensor itself emits a colored light the intensity of which decreases when it binds with the glucose molecule. Reduction in the intensity of the colored light, in such cases, is proportional to the glucose concentration and may be calibrated in terms of glucose concentration.

(iii) **Other Methods**:

Human blood is made up of four different components: (i) **blood plasma**, the fluid in which other components of the blood reside. Blood plasma contains 91–92% water, some salts like sodium (Na), potassium (K), etc., and antibodies, clotting agents and proteins like albumin and fibrinogen. (ii) **Red blood cells** (iii) **white blood cells** and (iv) **Platelets**.

It has been observed that an increase in the local glucose level in blood results in a decrease in the sodium level and an increase in the potassium level of the blood plasma. This changes the electrical properties, like the conductivity, dielectric strength, etc., of the plasma. The degree to which these electrical properties get changed depends on the glucose level and may be used to estimate the blood plasma glucose concentration.

10.2 Self-monitoring Glucometer (SMG)

Portable/handheld or self-monitoring glucose meter (SMG) is the device that is frequently used by patients for measuring their blood glucose level every day or few times a day. A SMG consists of three basic components.

(a) A ***lancet***, which is a small, sharp and fine needle. The needle normally resides in a protective plastic cover and can be pushed out with some force by pressing a spring-loaded button. The lancet, when put at the tip of a finger and the spring-loaded button is pressed, the needle jumps out making a small prick at the fingertip. The finger may then be squeezed to get a drop of blood. A lancet is shown in Fig. 10.2c.

(b) **Meter** is a small plastic box. Meter (see Fig. 10.2b) houses, a battery, a miniature but powerful computer and other microchip based electronic units like microcurrent meter, amplifiers, analog-to-digital converter (ADC), a liquid crystal display screen, etc. There are slots on one side of the meter in which a special strip or sensor may be inserted.

(c) **Glucometer sensor** or **Test Strip** that is specific for each instrument. The strip, in most cases, uses an electro enzyme approach. There are three conducting electrodes on the strip (made by depositing a thin layer of some conducting material) as shown in Fig. 10.2b. The working electrode on the strip is painted with a glucose ***oxidase enzyme*** which acts as a catalyst. There is also a very fine capillary on this electrode that sucks a fixed volume of blood from the blood drop on the fingertip when brought in contact with the strip. When the fixed volume of blood in capillary (with glucose) comes in contact with the oxidizing enzyme, the glucose contained in the blood gets oxidized by taking oxygen from

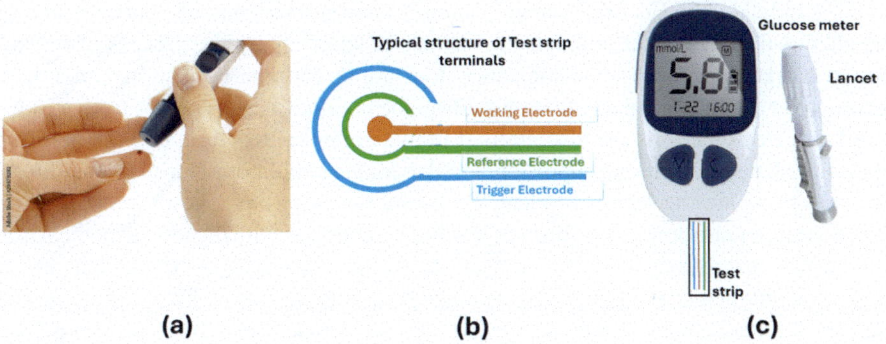

Fig. 10.2 **a** View of fingertip pricking by a lancet to draw a drop of blood, **b** the sketch of the three electrodes on the test strip, **c** a view of the glucometer with strip and a lancet

the enzyme. The oxidized glucose releases electrons on the working electrode constituting a current through it. The magnitude of the current is proportional to the amount of glucose in that volume of the blood. The generated current signal is converted into a voltage signal using the electronic circuits in the device. Conversion of current into voltage is required for further processing of the voltage signal, its amplification and digitization. The voltage signal is digitized and calibrated by putting glucose samples of known concentration on the working electrode and recording the generated digital voltage. The calibration data (digital voltage value and corresponding glucose level) is stored in the memory of the computer. The voltage signal from the blood sample of an unknown patient is then converted into corresponding glucose level using the calibration data either in mmol/L or mg/dL unit, as required.

As already mentioned, most SMG works by analyzing a small amount of blood, a drop only, usually taken from a fingertip. A lancet (tiny and fine needle) lightly pricks the tip of a finger of the patient from where a drop of blood is squeezed out and put directly on a special test strip. The test strip is inserted in slots provided for in the meter. A reproducible (or fixed) amount of blood from the blood drop put on the strip is sucked by a fine capillary connected to the working electrode. Since the working electrode on the strip is painted with a chemical called ***Glucose Oxidase Enzyme*** (often denoted by GOx or GOD, chemical symbol $((C_6H_{10}O_5)_n)$, the chemical enzyme catalyzes the chemical reaction of glucose (of blood) with oxygen to produce two new chemicals as indicated by the following reaction:

$$\underset{\text{(from blood)}}{\text{Glucose}} + \underset{\text{(from surroundings)}}{O_2} \xrightarrow{\text{Glucose Oxidizing catalyst}} \text{Gluconic acid} + H_2O_2 + \text{Free electrons} \quad (10.2)$$

As is indicated by Eq. 10.2, enzyme glucose oxidase does not take part directly in the chemical reaction but indirectly facilitates in speeding up the process of oxidation of blood glucose. When a substance does not directly participate in a chemical reaction but facilitates the reaction, the substance is called a catalyst. It is further indicated that free electrons are also emitted in this chemical reaction, and the number of emitted free electrons is proportional to the concentration of glucose in the blood. Free electrons are the carrier of electric current, the electronic circuits contained in the glucometer and connected through the three electrodes with the test strip, measure the electric current generated by the free electrons released by the blood sample. A larger current means a higher glucose concentration in the blood. The current generated by the free electrons released in the chemical reaction is first converted into voltage, and the voltage signal is then amplified and digitized. Conversion of current into voltage is required for further processing of the signal. The digital voltage values are calibrated in terms of glucose concentration using samples with known glucose concentration. The calibration data of output digital voltage and glucose concentration is stored in the memory of the computer, inbuilt in the glucometer, by the manufacturing company that carries out the calibration experiment before putting the device for marketing. This calibration data is used to determine the glucose level of an unknown blood sample from a patient in subsequent measurements.

Figure 10.2b shows the structure of the test strip. There are three conducting metallic coatings on the strip, called electrodes. Out of the three electrodes only one, called working electrode, remains projected out with a fixed volume capillary, to hold the blood sample from the patient. A fine capillary, an extremely narrow tube, has the property of pulling blood into itself. This occurs due to the property of ***surface tension*** of blood. Without going into further details of surface tension, it may be said that all liquids possess the property of surface tension. Since the volume of the capillary is fixed, each time the same volume of blood (from the blood drop) is analyzed for sugar (glucose) concentration. The other two electrodes, the reference and the trigger electrodes are required to provide appropriate electric voltages to run the electronic circuits inside the meter.

In short, when a fixed volume of blood sample is placed over the working electrode of the test strip, a chemical reaction takes place that generates a small electric current proportional to the concentration of glucose in the blood sample. The current is constantly monitored while the test strip is in place in the slots of the meter, allowing the device to register when the blood is put on the strip. The current is then converted into voltage which is digitized. The chemical reaction taking place at the working electrode stabilizes in about 5 seconds, after which the digital voltage generated by the chemical reaction is monitored by the device and is converted into corresponding glucose concentration using the calibration data already stored in computer memory.

10.3 HbA1C Sugar Test (Glycosylated Hemoglobin Test)

Glucose is a type of sugar in human blood that comes from the food we eat. It sticks to hemoglobin, which is a protein in red blood cells (*Blood is a specialized body fluid, it has four main components: red blood cells, white blood cells, plasma and platelets. Red blood cells contain a special protein called hemoglobin, which helps carry oxygen from the lungs to the rest of the body and then returns carbon dioxide from different parts of the body to the lungs so it can be exhaled*). As the glucose level in blood increases, more hemoglobin molecules get coated with glucose. The average life of a hemoglobin molecule is around three months. In an HbA1C glucose test, the number of hemoglobin molecules that are coated with glucose (called glycosylated or glycated hemoglobin) and those not coated with sugar, in a given volume of blood, is determined. Using this data, the percentage of glucose-coated hemoglobin molecules is determined. The result of the test is generally presented in percentage (%). Since the average life of hemoglobin molecules is around three months, the HbA1C result gives the percentage of glucose coated hemoglobin molecules averaged over a period of two to three months. It may be noted that the HbA1C test is different from a finger-prick test, which is a snapshot of blood sugar level at a particular time, on a particular day.

A common definition and method for HbA1C determination were developed by the International Federation of Clinical Chemistry and Laboratory Medicine (IFCC). The HbA1C test is mostly carried out in approved laboratories using modern techniques of chemical analysis (like high-performance liquid chromatography, electrospray mass spectroscopy, etc.) that are beyond the scope of this discussion. The test is a biochemical parameter used to examine the long-term control of sugar level in blood. The normal values of HbA1C range from 4.0 to 6.0%. HbA1C value below 5.7% characterizes the non-diabetic range; values in the range 5.7–6.4% are considered as prediabetic (i.e., high probability of becoming diabetic unless strict control on dietary habits and exercise regime is done). The value of the parameter above 6.5% indicates a diabetic patent.

As indicated, conventionally the HbA1C level is measured in percentage (%); however recently it may be given into two new units, mmol/mol (millimole per mol) and mmol/L (millimole per liter). Corresponding values in the old and the new units for some cases are given below (Table 10.1).

Table 10.1 Glycated hemoglobin level in different units

Old HbA1C value (%)	Value in mmol/mol	Value in mmol/L
6.0	42	7.0
6.5	48	7.8
7.0	53	8.6
7.5	58	9.4
8.0	64	10.2
9.0	75	11.8
10.0	86	13.4

Index

R. Prasad, *Understanding Medical Devices*, Series in BioEngineering,
https://doi.org/10.1007/978-3-031-88630-0

The manufacturer's authorised representative in the EU is Springer Nature Customer Service Centre GmbH, Europaplatz 3, 69115 Heidelberg, Germany. If you have any concerns regarding our products, please contact ProductSafety@springernature.com

Printed and bound by CPI Group (UK) Ltd, Croydon, CR0 4YY

05/10/2026

02228157-0001